Gravity is Just . . .
That Electrons are a Little Closer

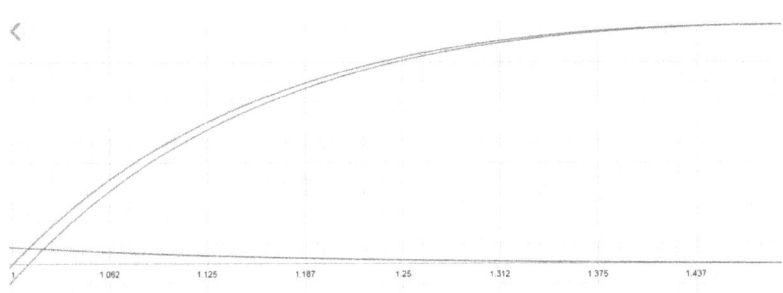

Explaining Gravity from the Basics of Electromagnetism

and

Explaining Why 'Mass' Changes

By Arno Vigen

© Arno Vigen, 2017/08

Simple Words to Understand . . . Atoms and Chemistry

Why does a Nucleus Stay Together If Protons Repel?

- A Nucleus is Just . . . a Nucleomagnetics Ring

Why Don't Electrons Fall into the Opposite-Charged Nucleus?

- Electrons are Just . . . Frightened by Nucleus nucleomagnetics

Electron Shell Chemistry Is Just . . . Scrunched Cube Geometry

- Why are electron shells in sets of 2, then 8, then 8 and such? Can we improve Pauli-aufbau?

Scrunched Cube Periodic Chart of Elements

- What are the properties and groupings of elements using the Arno Vigen Scrunched Cube model

Scrunched Cube Chemical Bonding

- Why does Nitrogen and Oxygen have Different Bonding Angles? What drives bonding?

What Makes a Molecule Solid, Liquid, or Gas?

- And Why is the Gas of Every Element the Same Volume (a mole)?

Simple Words to Understand . . . Gravity, Electromagnetism, and Other Forces

Gravity is Just . . . That Electrons are a Little Closer

- Explaining Gravity from the basics of Electromagnetism and Explaining Why Observed Mass Changes

Does Time and Space Really Warp?

- Replacing Electron-Shell Radius for Time-Space Factors in formulas such as the General Theory of Relativity

How are Electricity and Magnetism Linked?

- Exploring the Fundamental Linkage of Charge and Magnetism

Restoring Newton and Fixing Einstein's $E = mc^2$

- Given mass $m = \frac{M(z,n)}{\frac{8}{3}\pi(R_{ES})^3}$ defined by AVSC nucleomagnetics, then what is mc^2

Beyond the Quantum Era

- Restoring Newton, and Moving Beyond Quantum Mechanics

Simple Words to Understand . . . Personality

Visual Astrology: Fun, Support, Security, and Growth

- Astrology 'signs' archetypes are based upon powerful traits to understand people

Visual Astrology Relationships

- What happens when 'sign' personalities interact

Visual Astrology and Jung

- Astrology 'signs' archetypes actually predict all the Jungian 4 archetypes

Dominant Personality Traits

- Dominant Personality Traits Follow from Four Dimensions, Six Steps and so 24 Subcategories

Simple Words to Understand . . . Communications

Decision Matrix® Writing

- Persuasion is based making arguments at the correct strength in a certain order.

GATESOUP® Writing

- **G**oal, **A**udience, **T**heme, Enough **E**lements, **S**upport and the rest

Kedarf® Grammar and Composition Explained

- Defining the Parts of Speech, Paragraph Structure and More in Usable Terms

Table of Contents

Simplifying Concepts 8

Fundamental Question: What is Gravity? 11

    Charge-Force is One Part of the Updated Gravity Equation 11

    Electromagnetic Charge is the top fact for all of Chemistry 12

    Charge has an Attraction Force 13

    1/Distance-Squared ($1/d2$) goes forever and decrease exponentially 14

    Electron-shells is Another Part of the Equation 15

    1/distance-squared makes closer items more powerful 16

    'Big' Charge versus Magnetics versus 'Tiny' Gravity 17

Fundamental Question: Is the Gravity Force Directional? 19

    Net Bonding Force (Gravity) Averages, so Gravity Seems Constant 23

Fundamental Question: The Revised Formula for Gravity 25

    Part 1 – Adding the Radius of the Electron Shell to the Distance-Squared Gravity Formula 25

    Part 2 – Finding a stable RES: 28

    Part 3 - Charge Field = Magnetic Field Substitution 29

Calculation of Gravity from Charge versus Newton/Cavendish $6.674×10^{-11}\ m3kg(s2)$ 36

Fundamental Question: What Causes Mass? 39

If the Universe is in Balance, How Can There Be Overall Net Attraction? 43

Newton Discovered Both Gravity and Integrals, and this Extension of his Gravity Formula Applies Both of Newton's Discoveries 46

Challenge: Theorem Says Gravity Gets Based upon Charge, but Charge does not equal Mass which in the equation? 50

Why Separated Mass (Protons plus Neutrons) and Charge of Just Closer Electrons Got Missed 52

Fundamental Question: Why does the magnetic field from the nucleus # of particles get used as 'mass' when there is also a charge field moving electrons involved? How did just protons own charge get excluded if this is a Charge calculation? 54

Challenge: How can magnetics which is 1/-distance cubed $(1/d3)$ be part of a 1/distance-squared $(1/d2)$? 56

Calculation of magnetics does impact atom-gravity in bonding and reactions – close distances 56

Challenge: But the Net Charge of Every Atom Equals Zero – the Protons Equal the Electrons? 58

EMP Example Shows When Differential Eliminated 58

Challenge: Do Electrons and Nucleus Orbit Each Other, So Center of Gravity Offsets AVSC Net-1/Distance-Squared Calculation? 59

Fundamental Question: Do Electrons really have Little 'Mass' or No 'Mass'? 60

Why does some research find a tiny mass for an electron? 61

Challenge: Why does 1s Shells keep getting closer (smaller radius) when elements are increasing the magnetic field? Shouldn't the 1s Shell get further away? 62

Fundamental Question: What Happens with Temperature? I see magnetics decrease, but gravity does not? 65

Fundamental Question: What Explains why Mass Changes in Bonding? 67

Bonded Electrons do not contribute to the external Atom-gravity 'mass' 68

Decrease offset in that Electron still contributes to push out other electrons 70

Decrease offset by increases in distance to electron in other atom in the bond 71

Must also realize location is not fixed, but a quantum path 73

Fundamental Question: What Explains why Mass Changes in various Einstein equations? 74

    Atoms approaching near speed of light build towards infinite 'mass' 75

    Part 2 – Electron Speed in Orbit More than Speed of Whole Atom... 77

Fundamental Question: If Gravity is an Integral over Time, then are there Gravity Waves? 79

    Can Gravity Waves Synch? 79

For the Daring 80

Background from Prior Books 85

Revised Gravity Constant Derivation based upon Charge and Electron Field 89

Explaining Why Electromagnetism was Missed as the Basis of Gravity 91

Gravity is Multiple Particles Forces from Different Atoms Combined 93

$e = mc2$ Using Nucleus Particles instead of Mass 94

Impact of Direct Calculation of Gravity from Electrodynamics on String Theory and Such 97

Calculating the Current point-distance Gravity into a Net-Charge Surface Integral Not via Field Strength Shortcuts 100

The Complexity is Calculating How Far Away the Electron-Shell Separation Hovers from the Nucleus 113

Arno Vigen Science Postulates: 118

Endnotes 126

Simplifying Concepts

Big hugs for everyone.

Real understanding makes life easy. Life is a breeze and joyous when "you have your priorities straight". All those rules like "you need 10,000 hours of practice to become an expert" all come down to those magic moments when lots of factors, which to most other people seem complex, and which you actually studied, suddenly have a clear order to you, and your life in that area become simple, and powerful, and joyous. It becomes your career, your passion, and your source of pride.

That is the key:

- One thing is most important for each particular career, each industry, each area of science, each personality, or each financial calculation.

- Other things have a priority order in which to apply them.

Many people get that understanding intuitively. Great! That is why there is the 10,000 hours 'rule' for intuitive people. Intuitive people need to experience all those pieces in dozens of different ways; after that, the most important thing and the order for others get engrained by experience. They become experts in the area of life upon which they focus enough hours.

Gravity is very intuitive. We live with gravity every day, and know how to deal with it. However, gravity is not the top priority. Most people do not need deeper understanding.

Yet, gravity is *the important factor* for building twenty-story cellular towers -- which I have done. Some of us engineer types are not very intuitive; understanding of importance, sequence, and priority must come explicitly. We need a plan, and a piece of paper. Hence, this presentation matches my style, more concrete and less intuitive. Sorry if intuitive types must struggle to follow.

So why struggle to read this. Well, sometimes, we are lucky that other people before us find this critical order, communicate it well, and that means that millions after them don't need to spend 10,000 hours for something important and useful. Understanding comes from a few hours reading Sir Isaac Newton and the Laws of Motion. Sir Isaac Newton and the formulations of calculus. <u>7 Habits of Highly Effective People</u>. <u>5 Languages of Love</u>. Adam Smith, Milton Friedman, and John Nash (<u>A Beautiful Mind</u>) for interpersonal economics. There was one of those perfect moments of understanding in that movie, <u>A Beautiful Mind</u>, when John Nash said, "Adam Smith needs updating," and the new, better order of priorities became clear to him.

We stand on the shoulders of giants. I am glad they each spent their 10,000 hours, and found a way to explain and to help the rest of us launch our own quests.

Of course, it is the struggle that makes the understanding powerful. That is why it sometimes takes 10,000 hours, and sometimes it only take two (2) hours struggling to understand how an expert got to their understanding.

The back half of this writing is an impossible struggle for most people. To get to that core, a person already needs the 10,000 hours before actually trying the factors and calculations.

However, the first half of this reading should be the breeze, or at least an interesting challenge. The book is meant that 90% of people read the first half, and do not need the deep calculations at the back. I will be happy if everyone would already get the entire concept understand reading pages 10-36, struggling only on page 25 and page 30, the two ugly math pages. The back half is only for those who like spreadsheets, footnotes, and endnotes. Stop when you understand; I won't be mad.

After this exploration, the concept of gravity is simple now that the most important thing, charge, and the priorities of the rest are organized. The concept solves some fundamental questions. That means you can understand gravity completely within an hour or two.

Hopefully, my way of looking at these challenges makes the unknown makes sense. I know that I think quite differently the most people, but as Robert Frost said, "I took the road less traveled, and that made all the difference."

Big hugs. Let's get to work.

Arno

Fundamental Question: What is Gravity?

Gravity is the electromagnetic charge attraction for electrons (-) in their shell positions which is just a tiny net 1/distance-squared closer for distant objects than the positive (+) charges from the nucleus particles.

Charge-Force is One Part of the Updated Gravity Equation

Charge as discussed here is basic electricity. It is very powerful, as the lighting experiments of Ben Franklin can attest.

Even tiny bursts of charge can blow up things and stop hearts unless properly understood and managed. Charge also lights our cities so bright that anyone can see electricity at work from deep outer space.

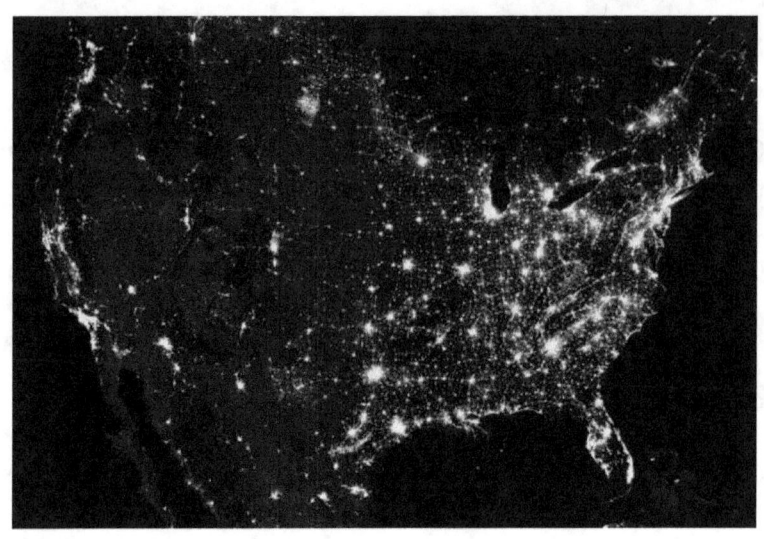

Every atom is a powerful balance of the positive charged protons and the negative charge electrons. For different matching numbers of positive and negative charges, all the periodic table properties for all the variety of elements differentiate. A diamond is a diamond because of its six protons and six electrons configuration. Iron is very magnetic because of 26 electrons in their configuration. All the element properties come from a complex structure starting with the Electromagnetic Charge force.

Electromagnetic Charge is the top fact for all of Chemistry[i]

Charge has an Attraction Force

Charge has a powerful attraction that we all know as well. It also has a repelling force. The world always comes in that sort of balance.

Charge-force follows the rule:

- Like Charges Repel

    o Positive (+) to positive (+) repels
    o Negative (-) to negative (-) repels

- Opposite Charges Attract

    o Positive (+) to negative (-) attracts
    o Negative (-) to positive (+) attracts

Charge is the most powerful force in the universe. It is the top of the chain. Everything starts with Charge – including gravity.

Charge attraction is governed by Coulomb's Law.

$$F = k_e \frac{q_1 q_2}{d^2}$$

q = charge of each molecule, and

k = the charge-strength constant.

Please focus that Charge attraction decreases at 1/distance-squared. This is similar to the current definition of the Gravity force.

$$F = G \frac{m_1 m_2}{d^2}, \quad G = 6.67 \times 10^{-11}$$

That common $1/d^2$ is the key, most important place to start. It makes sense that Electrostatic Charge-Force and Gravity-Force are directly related. Both decrease at the same rate. The evidence is hugely documented and proven that both charge and gravity act in 360 degrees, even spherically, in exactly the same ways.

1/Distance-Squared ($1/d^2$) goes forever and decrease exponentially

1/distance-squared in a formula means **a) that forces goes on forever** (and that the distant object looks at that force from one of those directions). The force gets smaller, but never disappears. Ah, like gravity! Charge (and gravity) never disappear. They keep working even a galaxy away. The square also means that **b) the change in distance reduces the force exponentially**. A 1/distance ($\frac{1}{d}$) calculation for a distance of 3 is three times (3x) as far, but 1/distance-squared ($\frac{1}{d^2}$) is not three times (3x) smaller, but 9x smaller. **Squared is exponentially more powerful.** Remember that a) and b) *double fact* for later.[ii]

My postulate is that gravity is a direct derivative of electrostatic charge. It is spherical; it decrease at 1/-distance-squared.[iii]

Now, the challenges is what makes gravity different . . . and smaller.

Electron-shells is Another Part of the Equation

The placement of electrons in shells creates a difference in Charge force that drives the gravity net-force, 2nd part, of the equation.

But, on the surface, this part of the theorem, that electrons are closer, seems challenging. An 'orbit' is both closer and further away. So, doesn't that cancel also? The back ½ cancels the front?

No, the close and far don't exactly cancel. A 1/distance-squared (*remember squared is exponentially more powerful*) force creates more than enough extra pull on the close side. The attraction at the close side of the orbit locations exceeds a) the repulsion from the protons in the nucleus; even with b) low attraction when the electrons are on the other side. That is the nature of 1/distance-squared.

A simplified example will explain why the nucleus-shell structure with electrons in 'orbit' generates a 'tiny' net force at 1/-distance squared:

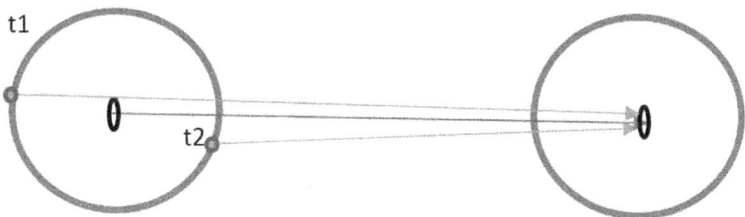

For simplicity[iv]: compare the force which acts using the distance-squared (as is the case for electromagnetic Charge over time, and thereby gravity) when the nucleus sits at a simple 10, say meters, (dropping all zero exponents/E-XX) and the electron moves in an orbit of 1.

That makes the nucleus each sit at 10 apart, and the electrons orbits at a distance from 11 at time1 and 9 at time2. The system moves back and forth 11 then 9, then 11, then 9.

Interaction / Time	Force / Distance-Squared	Calculation	Subtotal (in Force units)	%
Proton<>Proton				
Time 1	$-1.0 / 10^2$	0.01000		
Time 2	$-1.0 / 10^2$	0.01000		
- Total			-0.02000	100%
Electron<>Proton				
Time 1	$1.0 / 9^2$	0.01234		
Time 2	$1.0 / 11^2$	0.00826		
- Total			+0.02060	103%
Grand Total			+0.00060	3%

So, whenever there is a nucleus-shell configuration, charge **must** create some net-over-time (as they orbit) force towards distant objects. It is not 3% because my example numbers do not reflect the actual shell-distance difference ratio, but there is some 'tiny' extra-net charge force that has not been included in previously calculations trying to determine gravity.

That net-charge goes on forever. That is the nature of 1/distance-squared. It has all the common properties of charge **AND** gravity.

1/distance-squared makes closer items more powerful

Obviously, we need to add a few zero's to get the correct calculation. However, any person can understand the basic concept of the above closer-electrons average-over-time 1/distance-squared calculation. There must be a 'tiny' extra force when two opposite Charges exist in a structure of nucleus and orbits. With the electrons pushed out, the outer force is net a little closer for any $1/d^2$ calculation which makes them slightly stronger, and that 'tiny' net-force is gravity.

'Big' Charge versus Magnetics versus 'Tiny' Gravity

Without going into details, you can see that atom-gravity is tiny compared to the Charge Force. For two 001-H Hydrogen atoms at 1-meter apart Charge of 10^{-28} is ~1,000,000,000,000,000,000,000,000,000,000,000x more powerful than the 10^{-62} force for gravity. Charge is nearer to 100x (10^{-2}) bigger than Magnetics. Gravity is a 'tiny' force when compared to a) the electrical charge of an atom or b) the magnetic field of the nucleus.

Factor for 001-H <> 001-H atoms@1m	Force Calculation	
Electrical Charge Force factor	10^{-28} m^1 / (s^2)	*
Magnetic Force factor, if both north>south aligned	10^{-38} m^1 / (s^2)	*
Established Newton Gravity Force-Earth	10^{-62} m^1 / (s^2)	*

See chapter 9 for specific calculation.

Common sense tells you this order, even if the units are different and beyond human comprehension.

In the smallest spaces of our world, if you have ever gotten a shock from an electrical cord or static electricity, then you know a very small, almost-invisible electrical charge is powerful. A tiny electric cord short-circuit can create visible reactions. Gravity has never given you that sort of shock from something that small. If you have every tried to pry apart two magnets connected north to south, you know that magnetic force is strong, especially those tiny magnetic Bucky balls. Yet, again, no shocks from magnetic balls. Compare this to the atom-gravity force holding atoms of water into a bubble on a flat surface; one touch and the 'tiny' atom-gravity force of the bubble gets burst. Atom-gravity falls apart easily. Atom-gravity is tiny.

For gravity to really impact your life, it must be a huge boulder accelerating over a long distance. Of course, even then, a lightning bolt is more explosive. In the big world, charge is still much bigger than gravity.

At this point, what my fellow explorer should understand is that for a nucleus-shell structure, there must be a tiny net-force of the external electron attraction. And by tiny, I am thinking of how small 10^{-62} is versus 10^{-28}, so we are probably on the interesting path.

Next, we will move to figure out a) what causes the reflection, then b) how to calculate it, then c) check that the new method of calculating matches Newton.

Fundamental Question: Is the Gravity Force Directional?

Yes, of course. This same extra pull of whether the electrons are closer is the most basic calculation of molecular bonding today. In bonding, the direction of electron-nucleus force orientation is the core element of bond angles. **Bonding works because the electrons are closer in some directions when facing another atom than other locations and angles. That relative strength makes a particular nucleomagnetics angle the favored place for one atom to settle given another atom of a particular Element nearby; the positioning creates the bond.**

Bonding force is directional. Bonding force is gravity. Gravity is just that same principle applied from every direction over time and from far away so the bonding attraction and repulsions average out; that average for distant objects being a slight attraction.

Yet, for planets millions of miles away, of course, the calculations are the averages – which with millions of atoms rotating do average out such that the force seems the same in every direction.

In math, an *integral* is just averaging those, so please don't worry when I use that word. If you see or hear integral, just say average.

That bonding force is very directional. Sometimes it attracts; sometimes it repels.

- When there is a relatively open path from the distant atom to the nucleus, then bonding happen. From that direction, the shell-vs-proton net force shows as **attractive**.

Side View

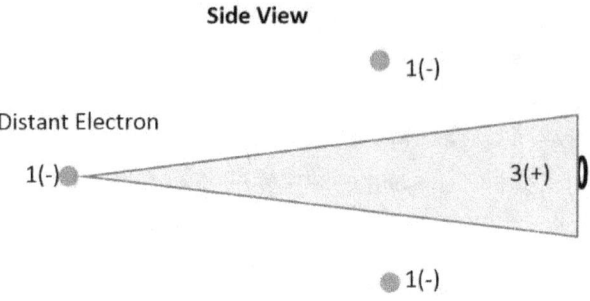

From this direction, the three (3) protons in the nucleus pull more than the two (2) electrons pushes sitting to the side of the nucleus. So, the shell-vs-proton net force **attracts**. *(By the way, the extra electron is on the far side, so please skip the complaint that I made an atom that does not have protons equal to neutrons.)*

- When there is an electron in the way, then the bond does not happen. From that direction, the shell-vs-proton net force shows as **repulsive**. The electron-shell is a barrier.

Side View – Electron Blocking Bonding Path

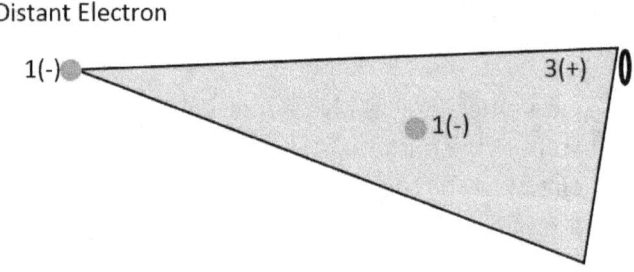

If the distant electron comes from a different direction, then the electron gets in the way, and the electron (-) to electron (-) repulsion overwhelms the attraction form the nucleus.

Remember that 1/distance-squared makes the closer items more powerful. As a result, at some point the distant electron gets repulsed by the closer, 1/distance-squared electron in the way.

As any atom gets closer, electrons first, then that ratio of the distance from electron shell versus the distance from the nucleus gets bigger. With 1-distance/squared, the force balance changes.

Distant Item Location	Nucleus Proton Distance	Electron Distance
3.0x Shell	3 − 0 = 3	3 − 1 = 2
2.0x Shell	2 − 0 = 2	2 − 1 = 1

Using those distances, the two charge calculations change from a net attraction at a distance (far away, like gravity!) to a net-repulsion near to the electron-shell.

This is the above example with one (1) electrons in the way of three (3) protons. You can see the electron create the 'shell' as anything getting close become repelled by the 1/distance-squared electron. The nucleus contains three (3) attractive to the one (1) electron repulsion.

Distant Item Location	Proton Pull	Electron Repel	Net
3.0x Shell	$3/(3^2)=0.333$	$1/(2^2)=-0.250$	+0.083 attracts
2.0x Shell	$3/(2^2)=0.750$	$1/(1^2)=-1.000$	-0.250 repels

Note that the close repulsion is bigger (-0.250) than the far away attraction (+0.083). If it comes any closer, this gets exponentially more repulsive. The attraction as a distance goes on forever. This net-attraction-from-a-distance, atom-gravity, even starts working at a distance only a couple times beyond the electron shell!

As the distant atom gets closer, if an electron is in the way, then it 'bounces' off the 'shell' of the atom. A set of electrons really is an invisible barrier, a 'shell'. In most directions, other atoms bounce off this invisible repulsion distance instead of bonding. Bonding is actually a very special event.

Also, remember that in everything but the simplest atoms (001-H Hydrogen and such), there will be multiple electrons in the way, and that the multiple shells are barriers (1/distance-squared).

Side View – Many Electrons Blocking Bonding Path

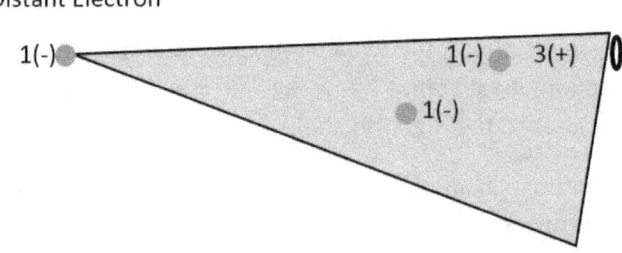

Directional Attractions and Repulsions

Therefore, the bonding attraction has this strength curve, both attractive and repulsive, around each atom.

Electron-Shell-vs-Proton Net Force By Angle

By Direction, Shell-Vs-Proton Net Force is Both + and –

The net-force is sometime positive, sometimes negative. Of course, the full picture is around the 3D sphere, but hopefully the 2D slice above gives the idea. Bonding force is all over the place – positive and negative.

Net Bonding Force (Gravity) Averages, so Gravity Seems Constant

In the world, all these various shell-vs-proton net forces are out there in different directions and the items themselves rotate:

- The Atoms Rotate so the Shell-Vs-Proton force goes up and down in the direction of the distant object;
- Different Atoms and molecules at any one time are oriented in different directions, further bringing the observation of distant objects to 'see' only averages;
- Gravity is so small that its impact is only over such enormous number of molecules so that only averages apply.
- Distance molecules have their own location and orientation. We only observed gravity when the distant object also contains a huge number of atoms.

Therefore, gravity is the combination of a bunch of little net-forces from many orientations. In math, adding lots of structured items is called an integral. Basically, an *integral* is averaging.

Gravity is the integral-average bonding force for millions of atoms applied for a distance object.

Fundamental Question: The Revised Formula for Gravity

Part 1 – Adding the Radius of the Electron Shell to the Distance-Squared Electrostatic Formula

The charges of the electrons are net-closer. How does that look in math terms?

Coulomb's Law

$$Force = \frac{Charge\ of\ Atoms}{d^2}$$

Before, picking a distant point and direction, a cleaner presentation of this is the field strength (that generates the force in a particular direction) which generally looks like:

$$Field\ Strength = \frac{Charge\ of\ Particles}{d^3}$$

[From a direction, d, a $\frac{1}{d^3}$ field generates a force for all the items in that direction, so the force is based upon $\frac{d}{d^3} = \frac{1}{d^2}$ as the factor ignoring other segments. There is -1/2 and such in conversion, but we are just working of the main factors in this section.]

I use field strength because that is a simpler calculation. I don't need the complexity of the extra directional distance calculation. I am taking a sphere of Proton Charge which decrease 1/distance-squared, yet comparing that to the sphere where the Electrons have not started their 1/distance-squared decrease until the electron-shell. The total sphere (field strength) is easier to calculate.

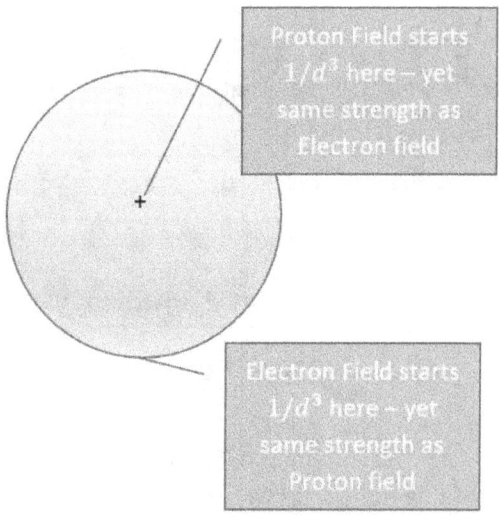

Proton Field starts $1/d^3$ here – yet same strength as Electron field

Electron Field starts $1/d^3$ here – yet same strength as Proton field

The total strength of both fields are exactly the same because protons equal electrons in all stable atoms. Both charge-force fields have the same # units and thereby, starting strength, but the electrons start decreasing later for distant objects by the Radius of the Electron Shell. Therefore, split into Protons and Electrons with the electrons closer by the Radius of the electron shell (R_{ES}):

$$\text{Distance}_E - \text{Distance}_P = (d - R_{ES})^3 - d^3$$

Expanding $(d - R_{ES})^3$, we can cancel the d^3 (and get the net field strength):

$$1/(d^3 + R_{ES}{}^3 - 3d^2R - 3dR^2 - d^3)$$

That extra $-3d^2R - 3dR^2$ in the denominator and a dx when you take the integral (average) from a distant point is a challenge. That explains NASA's problem with Mercury's orbit, but let's ignore here, it is a tiny tail 10^{-12} or 0.00000000001 smaller than $+1/R_{ES}{}^3$. Only the hearty will tread into this tiny tail (T) of a tiny net, and we will revisit the challenge of calculating with those factors in the final chapter. I look forward to vigorous debate as there are interesting choices in the dR and dx factors not discussed here.

Ignoring and removing the dR factors, the d^3 further cancel

$$\cancel{d^3} + R_{ES}{}^3 \cancel{-d^3}$$

So, we end up with the extra $R_{ES}{}^3$ which ends up in back in the formula as $\frac{1}{R_{ES}{}^3}$. Therefore, the net-field (electrons more than protons) has a $\frac{1}{d^3 R_{ES}{}^3}$ field strength factor, which as force becomes $\frac{d}{d^3 R_{ES}{}^3}$ factor which becomes $\frac{1}{d^2 R_{ES}{}^2}$ Gravity Force denominator from a particular direction for a distant object (again, ignoring some minor math details).

That main factor $R_{ES}{}^3$ makes connection of Charge to Gravity clear, simple and direct. Gravity is the leftover net-charge, and we can calculate it. **Gravity is no longer based upon a separate particle. It is clearly, intricately linked to the known electromagnetic Charge by a known factor R_{ES}.**

Part 2 – Finding a stable R_{ES}:

However, there is a challenge. The radius of the electron-shell varies. It is not based upon Charge (protons only), but the next magic is that the radius of electron-shell' radius (R_{ES}) gets directly based upon the magnetic particles (protons + neutrons) in the nucleus. **That shell's volume (R_{ES}^3) versus the nucleus magnetic particles averages *to the same ratio* no matter the number of particles.** That is because each magnetic nuclear particle delivers a fixed field (volume) strength of push. Therefore, the total push is the same, and the average radius of that pushed-sphere is a direct 1:1 radio.

From 1 proton, from 1 neutron, from 1 proton, and so on

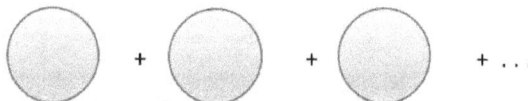

Every particle added in a nucleus add a certain volume of electron push, as a magnetic field. As a result, the R_{ES} steps in exact increments with the # of magnetic nucleus particles (protons, neutrons). It may have shapes (shells), but those are unimportant from a distant distance for the gravity calculation as we only use averages.

We see this with the elements, the radius of each element's outer shell expands. This has even been tested between isotopes of the same element. The differences between isotopes C12 and C14 are well known.

Part 3 - Charge Field = Magnetic Field Substitution

We know that the electrons 'orbit' at the balance of two forces – charge attraction and a repulsion just described. Therefore, we can use that for a substitution.

Here is the formula for Electrons Shells field strength balance point:

$$\frac{Attrative\ Net\ Charge}{R_{ES}^3} - \frac{Repulsive\ Magnetic\ Charge}{R_{ES}^3} = 0$$

so

$$\frac{Attrative\ Net\ Charge}{R_{ES}^3} = \frac{Repulsive\ Magnetic\ Charge}{R_{ES}^3}$$

or

$$\frac{Repulsive\ Magnetic\ Charge}{Attrative\ Net\ Charge} = 1$$

Note: This does not include d or distant objects. We are working only between in the field between the nucleus and the electrons shell for this Part 3 section.

There is a balance point of field strength and a stable place where electrons attractions matches nucleus repulsion *(when averaged over time and direction for those quantum fanatics).*

> The balancing point of the electron radius R is the most basic calculation of electron placement.
>
> R for various shells, without the normalized constant, is used in spectrum analysis for MRI's by knowing the wavelength in consistent, measurable R lengths from which photons are expelled.
>
> The energy to ionize (move an electron R length) is used in bonding energy. The distance R requires a consistent energy to move it to another orbit position.
>
> While those are different, they average to the same strength and volume of spherical energy field creating by N nucleus particles.

However, we do not need each electron's placement. We only use the overall balancing point.[v] Charge (that is, the number of proton (+) charge only) do not generate a universal* Constant. When the # protons change, they change R, but not in ratio. However, R_{ES} is very consistent compared to total nuclear particles (protons + neutrons). Nuclear particles (protons + neutrons) creates a universal* R_{ES}.

*'Universal' will still have various Einstein, Lorentz adjustments as discussed, integrated, and potentially improved in later chapters.

Therefore, in effect, Newton did not have knowledge of the left side and only used the right side of the below. [The electrons was not even discovered for another 100 years after Newton.] The effect was to replace the net-charge with a fixed-by-magnetic-particles calculation. Both are the same, but one moves in a 1:1 ratio.

$$\frac{\cancel{\text{Attrative Net Charg}}}{\cancel{R_{ES}}} = \frac{\text{Repulsive Magnetic Charge}}{R_{ES}^3} = \frac{\sqrt{G}(m_1)}{1}$$

And then that applies from two ends, so the Newton Theorem is:

$$F = \frac{\sqrt{G}(m_1) * \sqrt{G}(m_1)}{d^2} = \frac{G^1 m_1 m_1}{d^2}$$

I know that splitting the gravity constant to apply it as the square root to each of the factors is novel. However, that is one of the tricks that will make the path to the revised calculation clear, and convert to a formula where know particles and their direct electromagnetic fields comprise the formula.

The derivative of that is the field strength:

$$Strength = K + \frac{-2G^1 m_1 m_2}{d^3}$$

The method here is to work at field strength so that the clear path from nucleus magnetic particles becomes the Gravity formula and the *less clear* factors of mass and G get replaced with the well-defined factors of 1) (Z,N) = # of nucleus particles (Z=protons, N=neutrons), 2) G_A or M_A for the fundamental magnetic field strength, and so the gravitational contribution, of those particles; and 3) R_{ES} for the universal, stable* radius of associated N electron shells:

Which is the same as an integral of the Net-Field Strength which is the reduction of the Electron Shell Field strength at the balance point (big section) + C which is that same field at starting its strength.

This consists of:

$(G_A N_1)$ as the basic magnetic push, and thereby gravitation contribution, from the number, N_1, of nucleus particles (protons and electrons combined).

d = distance between the atoms

$\frac{4}{3}\pi R_{ES}^3$ for the volume of field strength sphere displaced by the electron shell of universal average radius R_{ES}*.

*'Universal' will still have various Einstein, Lorentz adjustments as discussed, integrated, and potentially improved in later chapters.

$$Strgth = \frac{-G_A^2 N_1 N_2}{d^3(\frac{4}{3}\pi R_{ES}^3)} = \frac{-(G_A N_1)(G_A N_2)}{d^3(\frac{4}{3}\pi R_{ES}^3 + T)}$$

$$Force = \int_{d-R}^{d+R} \int \int \frac{-(G_A N_1)(G_A N_2)}{d^3(\frac{4}{3}\pi R_{ES}^3 + T)}$$

Using the basic integral formula $\int \frac{1}{d^3} = \frac{-1}{2d^2} + C$

$$F = \frac{-(-(G_A N_1)(G_A N_2))}{2d^2(\frac{4}{3}\pi R_{ES}^3 + T)} + C = \frac{(G_A N_1)(G_A N_2)}{d^2(\frac{8}{3}\pi R_{ES}^3 + 2T)} + C$$

And the constant from which is starts is the field strength:

$$C = Constant = \frac{M_A(Z,N)_1 M_A(Z,N)_2}{(\frac{4}{3}\pi R_{ES}^3 + T)}$$

Where $R_{ES}>0$, of course, given that we will often ignore T, the denominator cannot get to zero. In real life, that cannot happened because an atom has a physical radius, so it is natural that $R_{ES}>0$.

That makes the charge force the netting of proton force and electron force with the electrons just that much closer (on average):

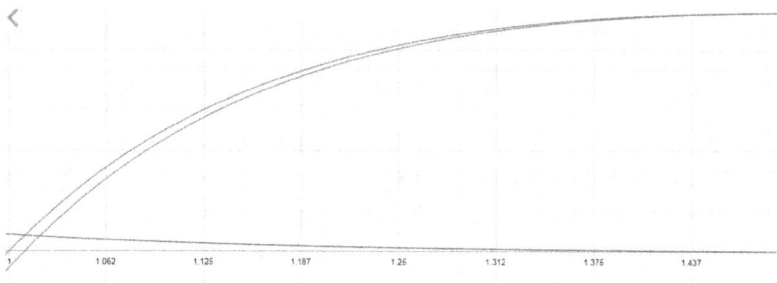

Atom-Gravity

Factor	Base	Field Strength decreases by	Radius
Nucleus creates magnetic field force	Protons + Neutrons = 'Nucleus Mass'	While complex to calculate specific electron balance points within a magnetic field[vi], the total field balance is simple: (Magnetic Repulsion = Charge Attraction) over / Volume of (Shell-Radius)³	Nucleus
Electrons has closer Charge force	Net Charge	$$\frac{kQ_E}{(d \mp r)^3} - \frac{kQ_P}{d^3}$$ Excess Charge of Electrons less the Charge of Protons averaged over time over the electron shell	Electron Shell**
Distant body recognizes only the 2nd larger element	Product with the Electron Shell's Charge as the 'math eliminated' connection	The Charge Attraction cancels to: $$\frac{M_A(Z,N)_1 M_A(Z,N)_2}{d^2 (\frac{8}{3}\pi R_{ES}^3)}$$ Which is the same as Newton $$\frac{Gm_1 m_2}{d^2}$$	Galaxies away

* I use Q versus $q_1 q_2$ because this is the fundamental of one Charge attraction or repulsion to one other Charge. It is not a variable, but 1:1.

** The electron shell distance is the combination of the electron-electron repulsion (spherical) plus the nucleus magnetic field (north-south oriented). However, for electron repulsion, there is the exactly balancing number of proton attraction, so in the end for the average the driving element is just the nucleus magnetic field.[vii]

Gravity by Arno

Nucleus particles magnetics (nucleomagnetics) cause electron-shell-distance

Nucleus Particle Magnetics Caused

Gravity is a relayed force.

Relayed

Gravity is a net force. The electrostatic charge-at-a-distance force of electrons is greater than a) Electrostatic Charge-at-a-distance force of the nucleus protons, even with b) lower Electrostatic Charge of electrons on the far side of any electron-orbits -- both because of 1/distance-squared.

Net-Charge-$1/d^2$ Force

Gravity is an integral over time. It is not a separate particle.

Derived *integral over time*

Gravity is a nucleus-particle-magnetics-caused, electron-shell-relayed, net-electrostatic charge-$1/d^2$-force calculated by an integral over time.

Calculation of Gravity from Charge versus Newton/Cavendish $6.674 \times 10^{-11} \frac{m^3}{kg(s^2)}$

These Electron-Shell Separation has many factors which will get handled extensively in the footnotes. There is a lot of moving parts, many of which need further validation for a 'published paper' level of acceptance. However, in general, you can see these implications of these factors in the below example of 001-H Hydrogen to Hydrogen at 1-meter apart:

- The Charge Force (if no offset by electron) is significant - 10^{-28}
- The Gravity Force (using Newton) is 'tiny' - 10^{-62}

- The Net-Electrostatic Charge Force (Arno Vigen method) is a similar 'tiny' - 10^{-62}

The Gross-Charge Calculation for 001-H Hydrogen electron (-) to 001-H Hydrogen (+) atom at a distance of 1-meter is as follows:

Factor	Net-Charge Calculation
Charge Force factor	$k = 10^{10}$ m² / (s²) [9.03)×10⁺⁹]
Charge of orbiting Electron	$Q = 10^{-19}$ [1.602)×10⁻¹⁹]
Charge of distance Proton	$Q = 10^{-19}$ [1.602)×10⁻¹⁹]
Distance	d = 1 m or 10^0
Exponent shortcut	+k+Q+Q-d-d
Gross Charge Short-cut calculation	10-19-19-0-0 = -28
Gross Charge Force	10^{-28} m¹ / (s²)

The Net-Charge Calculation for 001-H Hydrogen atom to 001-H Hydrogen atom at a distance of 1-meter is as follows:

$$\frac{M_A(Z,N)_1 M_A(Z,N)_2}{d^2(\frac{8}{3}\pi R_{ES}^3)}$$

Factor	Net-Charge Calculation
Charge Force factor	k = 10^{10} m² / (s²) [9.03)×10^{-9}]
Charge of orbiting Electron	Q=10^{-19} [1.602)×10^{-19}]
Charge of distance Proton	Q=10^{-19} [1.602)×10^{-19}]
Distance	d=1 m
Radius of the Nucleus (protons and neutrons)	r=10^{-11} m (Bohr radius which is 5.27 x 10^{-11})
Exponent shortcut	+k+Q+Q-d-d
Gross Charge Short-cut calculation	10-19-19-0-0 = -28
Net Electron-Shell Charge Short-cut calculation	-28-11-11-11-1 = -61
Net-Charge Force of the Electron (protons and neutrons) *New Formula*	10^{-61} m¹ / (s²) /8/3π = 10^{-62} m¹ / (s²)

This is just the basic force of charge force as pushed out by the electron-shell. The gross charge of the electrons 10^{-28} gets reduced by the $\frac{8}{3}\pi R_{es}^3$, which is another 10^{-34} creating a net-charge for distant objects of 10^{-62}.

How does that compare with the standard calculation of gravity using Newton's formula?

Gravity from 001-H Hydrogen atom to 001-H Hydrogen atom at a distance of 1-meter using the Newton method is:

Factor	Gravity Estimation
Gravity Constant (Newton)	$G = 10^{-10}$ m^3 kg^2/ (s^2) [6.67]×10^{-11}]
Mass of close 001-H Hydrogen	m=10^{-26} kg [1.602]×10^{-24}]
Mass distant 001-H Hydrogen	m=10^{-26} kg [1.602]×10^{-24}]
Distance	d=1 m or 10^0
Exponent shortcut	+G+m+m-d-d
Short-cut calculation	-10-26-26+0+0=-62+0=-62
Newton Gravity Force	10^{-62} m^1 / (s^2)

The new calculation gets to the same range of strength; net-charge at electrons shells matches Newton's gravity force for a simple 001-H at 10^{-62} m^1 / (s^2).

Net-Charge Force	10^{-62} m^1 / (s^2)

Fundamental Question: What Causes Mass?

Mass is the observed result, in a fundamental sense, the strength of the magnets of the particles, the protons and neutrons, in the nucleus. That nucleus magnetic field strength creates the push of the electrons into shells. That distance created by the magnetic field then converts into electrons further out, which in turn causes a stronger charge differential in spreading the location of protons versus electrons. That magnetic field creates a difference in the electron charge separation which makes 'mass' apply.

Mass as observed itself has no charge because the electron shell absorbs that energy. Mass = Magnetism usually expressed as net of the Electron Radius.

$$mass = \sqrt{G} \quad m = \frac{G_A(Z,N)_1}{\frac{4}{3}\pi(R_{es})^3 + T} = \frac{M(Z,N)_1}{\frac{4}{3}\pi(R_{es})^3 + T}$$

$$G_A = M$$

m = traditional mass

M = Magnetic Field Strength of nucleus particles

N_1 = Nucleus particles (protons + neutrons)

R_{es} = Radius of Electron Shell

G_A = Gravitational Field Strength of nucleus particles

The magnetics of the nucleus is not readily observed because that field is directional and decreases faster. Therefore, measurements are inconsistent. Further, the structure of nucleus changes the 'bagel' structure and relative strengths of the magnetic field. A shell radius is different lengths depending on where the electron rests in that magnetic field.

While the proton attraction is the same, the magnetic field strength and therefore repulsion varies by the direction (a bagel shape).

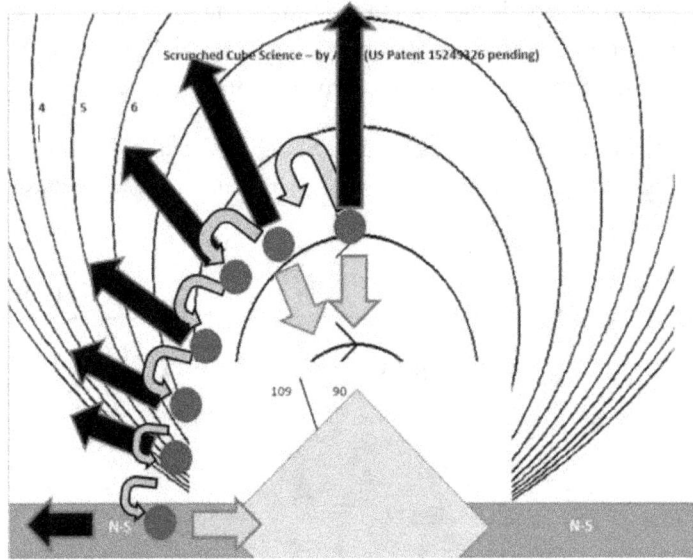

Note that when an EMP from a nuclear explosion temporarily eliminates the electrons, the underlying mass-(charge + magnetic)-force shows, and that wave is enormous and extremely destructive.

The magic is that mass is directly observed only as its reflection in charge at the electron shell radius. Do you know why?

Charge is the most powerful force in the universe. Even a reflection of magnetism in charge at the electron shell radius is stronger than the magnetism itself, at a distance, of course. That is how powerful Charge's 1/distance-squared works versus 1/distance-cubed.

The regular magnetism of the nucleus particles decreases at 1/distance-cubed:

$$\frac{M}{d^3}$$

Because the north-south pattern reduces faster than the spherical Charge. Only the relay to Charge does on further decrease by only 1/distance-squared:

$$\frac{kQ_1Q_2}{d^2}$$

Further, any direct calculation of magnetism is much more complex given the field is oriented north-south. It is not the perfect sphere of Charge. I treat this electron-shell balance issue more thoroughly in my Simple Words to Understand . . . Chemistry - Book 2, Why Don't Electrons (-) Fall into the Opposite Charge Nucleus Protons (+): Electrons are Just . . . Frightened of Magnets. It explains why electrons don't (-) just fall into the positive (+) charge nucleus.

That magnetism also gets further diluted and twisted by the structure of the nucleus itself. I treat the nucleus structure issue more thoroughly in my Simple Words to Understand . . .

Chemistry - Book 1, <u>Why Does the Nucleus Hold Together when Protons Repel Each Other: Nucleus is Just . . . a Magnetic Ring</u>. It explains that magnetism holds the nucleus together.

While working here in overall averages, the placement of electrons with a magnetics calculation actually improves understanding of the periodic chart and improves the aufbau 1s/2s/2d to resolve know inconsistencies. I treat the periodic chart structure issues more thoroughly in my Simple Words to Understand . . . Chemistry - Book 3, <u>Electron Shells are Just . . . Scrunched Cube Geometry: Why are Electrons Shells 2, 8, 8, and such?</u>

If the Universe is in Balance, How Can There Be Overall Net Attraction?

"For every action, there is an equal and opposite reaction."

Sir Isaac Newton

You have all these net electron-shell attractions going on forever. If so, how can it be that the universe is all these attraction forces? What is the offset? Where are the balancing repulsions?

The forces do offset, but in a creative way. The force does net over all distances from the atom. However, there is a huge negative net force, a tendency for repulsion, at the shells, and thereby a leftover attractive force outside. But that leftover goes on forever (although it gets really small), so the gravity-force actually balances the electron-shell repulsion zone. However, they are opposite types – big, fat at the shells versus long and skinny everywhere else.

The shell of electrons creates this huge barrier. We see this in all gases; we see this in the limited locations of bonding and bonding angles and strengths of bonds. An atom of a particular elements is highly protected by its electrons shell. When most other atoms approach, the other atom's electrons near the sitting atom's electrons first, and the atoms bounce off without any bonding.

As a result, there is a huge ring of repulsion because if other atoms, with electrons also on the outside, then as they get close, the electron repulsion overwhelms the general attraction. Another atom bounces away. It is rare and special to approach correctly to build an atom-to-atom bond that lasts.

Here is an example of the net attraction/repulsion for a 006-C Carbon atom based upon distance – in the line of one of the surrounding electrons. It has a tiny net attraction after the distance of about 2x Shell-2. It has a huge repulsion when it gets anywhere near Shell-1.

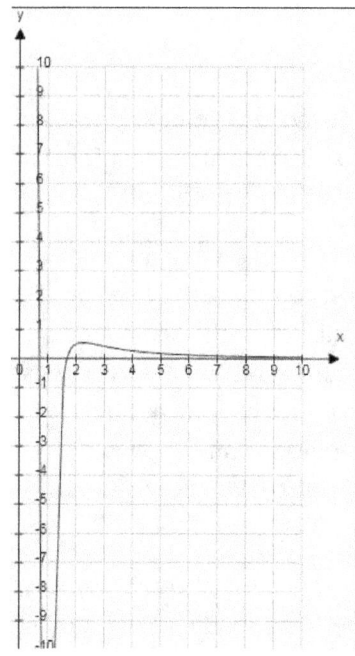

$$y = 6 \cdot \left(\frac{1}{x}\right) \cdot \left(\frac{1}{x}\right) - \left(\frac{1}{x-1}\right) \cdot \left(\frac{1}{x-1}\right)$$

Near Shell-1, the repulsion (negative) is enormous.

Attraction peaks at just over twice the distance of the electron shell.

Then attraction reduces – by 1/distance-squared, of course. There it is again, gravity is that tiny net left over, even if electrons are in the way. Gravity is something different near atoms; it is called bonding force and it is more often negative up close, and attractive far away.

At long distances, that large nucleus attraction in a direct line overcomes the electron repulsions which are distributed over the shells.

At short distances, the electron repulsions overcome the nucleus pull and that protects the stability of atom structure.

Over all distances, the world order balances. Strong repulsion nearby balances tiny attractions from distances going on forever. The two balance.

Newton Discovered Both Gravity and Integrals, and this Extension of his Gravity Formula Applies Both of Newton's Discoveries

It is amazing and funny that Sir Isaac Newton was both discoverer of gravity and inventor of the calculus of integrals, but Newton never had to apply the rule of integrals to his calculation of the force of gravity. (Electrons were not even observed and documented until Rutherford, Bohr, and, of course, Faraday a hundred years later. I trust Newton would have figured it out this discovery immediately.)

A gravity-orbit only has the one force with both objects locked into each other, so the integral gets the same result; it is easy to average. Scientist for a century have probably missed the connection.

Why Missed?

When the two gravitational-orbit around each other, the integrals are not very helpful for two reasons.

First, the mass, and thereby force, at each end is homogeneous. As such, there is not that huge different of (+) at the center vs the (-) distributed at the shell radius in the integral. When the shell is (-) that difference become important and measurable. For a planet, the whole distant object is homogeneous. For a nucleus-shell structure, the force is not homogeneous; the charge has opposite signs, positive at the center and negative over the shell surface area. Even with that, this calculation generates only a tiny difference ratio, 10^{-37}, so most people would not make the connection.

> Second, the integral for homogenous masses/forces gets back to an ellipse orbit because the two items are the only two interacting. However, for electron orbits and long-distance objects, there is more than just the gravity interaction, so the extra effort for integral finds the calculation. The shells create a 3rd factor that is oriented north-south, so it is not easy to measure.
>
> The gravity-orbit mass-at-a-point-distant methods works well enough. Only NASA needs to worry about some tiny variances that would make a spaceship miss Juniper with just is miscalculation of 10^{-11}. By the way, NASA has to deal with the homogeneous much smaller distribution of gravity objects at <10^{-10} accuracy or they miss the planet.

In the end, all I really do in this proof is change the long established Newton gravity formula:

$$F = G \frac{m_1 m_2}{r^2}, \quad G = 6.67 \times 10^{-11}$$

This calculation is easy because it assumes that all the mass sits at a point at a certain distance (r) from the distant location. There is no integral; that makes the calculation easy, but misses that fine detail.

For a charge distributed over 1-unit block at distance d, the Newton Gravity formula changes into an integral of the strength field:

$$F = \int_{x=r-1\,Planck}^{x=r} -2G_N \frac{m_1 m_2}{x^3} + C$$

Let's look at the elements for common sense.

The negative sign makes sense. As the force goes away, the force of gravity goes down. The total force is positive, but that force decreases.

The $1/r^3$ also now makes much more sense. The force spreads out evenly in three dimensions (X, Y, Z), and therefore, it reduces in exactly in the same three dimensions. If you take a sphere of radius 2, the volume of energy enclosed is $A = \pi r^3 = \pi 2^3 = 8\pi$, so the force at any point is $= 1/8\pi$. If that radius expands to 4, then the volume expands similarly $A = \pi r^3 = \pi 4^3 = 64\pi$, so the force at any point is $= 1/64\pi$. The total energy both remain 1 $(1/8\pi)(8\pi) = 1$ and $(1/64\pi)(64\pi) = 1$

The $1/d^2$ force in the Newton equation now makes sense. Only two of the dimensions get weaker relative to a particular direction. The X direction force adds up, so that direction all the pieces are still pulling the distant object. It really is d/d^3 view which becomes $1/r^2$. It has points along the line so only the ones in two (2) dimensions are not pulling the distant object.

The two formulas are exactly the same with just the addition step of doing the calculation for multiple points. The above formula assumes that all the charge sits at a cube with a diameter of 1.

It is as a bunch of little Newton gravity calculations for every combination of particles:

- Electron-to-electron,
- Proton-to-proton,
- Electron-to-proton, and
- Proton-to-electron.

I love Newton. This formula is actually moving back to Newton, which many current alternatives seem to question by adding non-Newton dimensions to get their results that work. As you review, you will see that using this definition, a number of 'extra dimensions' become a knowable force calculated from the Constant R_{es} (Radius-Electron-Shell) and the G_A or $\sqrt{G_A}$ knowing the secret that $\sqrt{G_A}$ actually equals the fundamental particle magnetism.

Let me be clear. Newton was 100% correct. My postulate is a method to apply *Newton* gravity as a *Newton* integral.

Challenge: Theorem Says Gravity Gets Based upon Charge, but Charge does not equal Mass which in the equation?

The charge of elements grows, but the mass grows differently in the below table:

Element	Charge	Atomic Mass	Ratio
001-H Hydrogen	1	1	1
002-He Helium	2	4	2
003-Li Lithium	3	6	2
004-Be Beryllium	4	9	2.25
005-B Boron	5	11	2.2
006-C Carbon	6	12	2
...			
026-Fe Iron	26	59	2.1
...			
074-W Tungsten	74	184	2.5

The challenge is that does not work if that is correct. The charge and the mass are not the same thing. There must be something that relates to mass in the formula.

However, by the 2-segments of the improved formula, separating net-charge from the magnetic-separation, we have separated into one element a factor that is based upon the magnetic particles of a nucleus. The new segment of the equation gets the 'mass' elements of the traditional linear portion of the equation. The second half is that charge which gets the 1/-distance-squared elements.

Instead, we are equating magnetic field with Atomic Mass. The postulate is the following chain:

- Sum of Protons and Neutrons > Magnetic Field

- Magnetic Field creates the 'electron-shell-distance-differential' (electrons are repulsed by magnetic fields)
- Electron Shell distance creates the net portion Charge to distant objects, but it is not the Charge itself.
- What those distance object see is a charge reflection of the sum of proton plus neutrons.

  :: Proton + Neutrons => Mass

Element	Protons + Neutrons	Atomic Mass	Ratio
001-H Hydrogen	1	1	1
002-He Helium	4	4	1
003-Li Lithium	6	6	1
004-Be Beryllium	9	9	1
005-B Boron	11	11	1
006-C Carbon	12	12	1
...			
026-Fe Iron	59	59	1
...			
074-W Tungsten	184	184	1

That table works great. Nucleus particles match nucleus-mass exactly.

This even goes to the isotopes which create similar elements, but with different mass. When you get an extra neutron, then by this theorem the magnetic field goes up, which increases the magnetic differential, and thereby, in ratio, the mass increases.

This is one of the first tests. The charge stays the same, but the magnetics changes (from isolated change of one neutron only). This isolates no change to charge, either the number of electrons or the number of protons, with a change to magnetics.

The more particles, protons and neutrons, that are in the nucleus; the more magnetic field strength that is created.[viii]

So, the argument that challenged scientist for years, and probably keep the revelation hidden is:

Why Separated Mass (Protons plus Neutrons) and Charge of Just Closer Electrons Got Missed

Prior solutions came to a serious impasse because the field strength follow charge, but the mass does not follow charge in an atom.

> Mass = Protons plus Neutrons
>
> but Neutrons do nothing with Charge

Calculation based upon charge did not include neutrons which never resolved.

Prior solutions came to a serious impasse because charge is involved in the solution, but the charge of an atom nets to zero.

> Charge > 1/distance squared and so is gravity

Charge must be involved, but net charge of atoms is zero. The basic zero times anything is zero made calculations based upon net charge fail.

Fundamental Question: Why does the magnetic field from the nucleus # of particles get used as 'mass' when there is also a charge field moving electrons involved? How did just protons own charge get excluded if this is a Charge calculation?

According to differential geometry mathematics, for example Gaussian random field calculations[ix], the strength of a field is the sum of the force elements enclosed. Once you get outside the space (sphere) where all forces are included, then that space is minimized based upon a simple sum of all those force vectors.

Gravity to distant objects is definitely outside the space of all the named elements (protons, neutrons, and electrons[x]). Gauss is a great shortcut.

So, for the electron field – as an average distance, yes, the three forces add, but two of them exactly offset. The positive charges of the protons in atoms are exactly equal in number, but opposite in Charge than the electrons. Therefore, for the differential geometry integral, those completely offset. That leaves only magnetics in the determination.

For an atom,

Protons create attraction	# of Protons
Electrons create repulsion	# of Electrons = # of Protons
Magnetic field keeps the electrons apart from the nucleus - repulsion	Based on protons plus neutrons (P+N)

The first two offset for the differential geometry vector analysis. This is a huge trick so you don't have to calculate the location of every electron in electron shells. Worse, you don't have to calculate the electron-to-electron repulsions for every permutation of electron combinations. [For a fuller presentation on actual electron shell placements see Book #2 of that Simple Words to Understand . . . Chemistry series on the shape and orientation of the electron-shells.] You can skip those complicated steps entirely.

Further, the full magnetic field is in direct ratio the sum of protons and neutrons in the nucleus. Therefore, it become a linear solution based upon nucleus particles (=mass).

So, that relationship between the electron distances is fairly consistent. The nucleus has XX particles and magnetic strength; therefore, the mass has the same ratio to those XX particles. We find that mass is directly related to the number of nucleus particles.

Most importantly, we find that the average distance of electrons in all shells is directly related to the XX nucleus particles.

The connection factor that those create the magnetic field strength is the missing lynchpin. The factor not addressed is the radius location of electrons given a specific magnetic source.

[I am still investigating if there is any electron feedback force to the magnetic field. So, this needs much more investigation. That I will ask for others to help investigate.]

[I am still investigating how the nature of neutrons includes protons subparticles, such as the neutron, and if how that adds magnetism because it has a proton inside. That I will ask for others to help investigate.]

Challenge: How can magnetics which is 1/-distance cubed ($1/d^3$) be part of a 1/distance-squared ($1/d^2$)?

That magnetic field become immaterial itself. It only helps create the separation. All the ($1/d^2$) force is done by the electron-differential charge. The magnetics impact only the other ½ for mass which is linear.

The magnetics becomes linear and not ($1/d^3$) because the separation is the average over the full sphere over time. That is the integral over every direction (3D) over time which brings the $\iiint (1/d^3)$ which is back to linear.

That the magnet pushes out the electrons means that 'mass' does not get based upon the half of the equation with **charge** 1/distance-squared ($1/d^2$) versus the 1/distance-cubed of quickly dissipating magnetic field ($1/d^3$). However, the magnetic field has done its job. It created the shell distance differential.

The magnetics is only in the 'mass' half which does not decrease over the long distances involved. That mass elements are linear, not 1/distance-squared ($1/d^2$).

Calculation of magnetics does impact atom-gravity in bonding and reactions – close distances

Also, not that any magnetic distance change is just the shell distance versus the distance object, so we probably could not object these.

I will address bonding and reactions in a separate volume. At those distances, the nucleomagnetics field is the important factor.

For gravity calculations, the magnetic elements are immaterial for calculations.

Challenge: But the Net Charge of Every Atom Equals Zero – the Protons Equal the Electrons?

Yes, that is why the result is tiny. The protons and electrons equal and offset each other. That is why the only thing left is a 'tiny' distance-difference for the just-closer-on-average-over-time electrons.

EMP Example Shows When Differential Eliminated

In the example of an electromagnetic pulse (EMP), the electrons are removed. At that time, the 'gravity' becomes massive. That is, in an electromagnetic pulse, the proton charge is alone and become gravity temporarily. The effects of gravity that is not 'net' are devastating to everything nearby. Remember the original comparison:

Factor for 001-H <> 001-H atoms@1m	Force Calculation	
Electrical Charge Force factor	10^{-28} $m^1 / (s^2)$	*
Magnetic Force factor, if both north>south aligned	10^{-38} $m^1 / (s^2)$	*
Established Newton Gravity Force-Earth	10^{-62} $m^1 / (s^2)$	*

Challenge: Do Electrons and Nucleus Orbit Each Other, So Center of Gravity Offsets AVSC Net-1/Distance-Squared Calculation?

No, the original theorem of Bohr-Rutherford stated that electrons orbit. And, yes, Newton has a calculation where the larger orbit also rotates so the center of gravity remains in the same place.

However, the Newton calculation requires that the inward force is offset by the angular momentum force. That is if the electron (or planet) were not in a generally perpendicular movement, the two would fall together. It only works if the gravitational orbit is at a consistent speed.

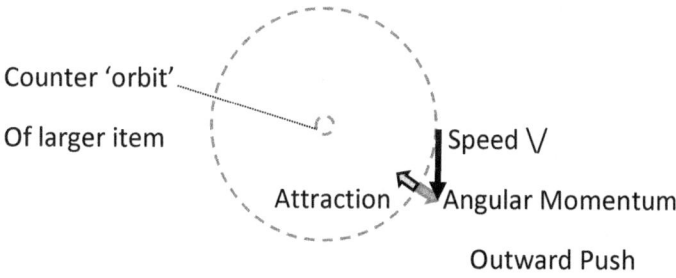

But electrons are not moving sideways. Electrons sit at a balance locations, and can wiggle in any direction based upon the relative forces of the nucleus electrostatic attraction, versus the nucleomagnetics repulsion plus any electron-electron repulsion from other shell positions.

The electrons do not need to move, and do not move, perpendicularly to stay at in the shell. There is no orbit, so this counter argument does not apply. The center does not counter-orbit.

Fundamental Question: Do Electrons really have Little 'Mass' or No 'Mass'?

Yes, but not in the way of current thinking. Mass is not some 'god particle' that is 1/1836 as small in electrons as it is in protons. Mass is not separate itself from nucleus particles. Mass is the combo-net-force described earlier. Gravity is derived, not inherent.

In a proton, the charge attracts, and the magnet when properly aligned also attracts. Therefore, one can measure 'mass' for a proton. Protons have elements of both parts of the gravity formula so they exhibit gravity behaviors that are measurable.

For an electron, a positive (+) charge attracts, BUT the magnetic field repels. In fact, it repels so well that electrons stay in some location 'shell.' Therefore, whenever you want to measure the 'mass' of an electron, it gets frightened and runs away. An electron only has the charge part of the charge segment of the gravity formula. It cannot exhibit mass or gravity because it does not have the other half, the magnetic component.

Electron Interaction

Particle Type	Charge Force	Magnetic Force	Net
Electron-Proton	Attractive	Repulsive	*Zero**
Electron-Electron	Repulsive	Repulsive	Measurable

If we are measuring the magnetic field repulsion distance, and electrons run away from any magnetic, then equipment have nothing to measure.

Further, today's equipment want to measure electrostatic energy and not magnetics.

Further, they do not know the nucleomagnetics axis orientation so they cannot get consistent magnetic field measurements.

Since 'mass' is the nucleomagnetics particles field, then electron which flee magnetic fields have no 'mass'.

Why does some research find a tiny mass for an electron?

The tiny measurement of electron mass that scientists do see of mass is the electrons get stopped by the field of the measurement devise. When it runs away, it stops the electron shell of the destination, not at the nucleus. Magically, the strength of the field at the shell is that same 1,836:1 ratio as the 'mass' of electron to proton ratio. The observed electron 'mass' is the reflection of the electron when it gets to the measurement atom and does not go all the way to collapse into that nucleus.

The mass of an electron, is the ratio of the diameter of nucleus to the radius where the average electrons sits (akin to the Bohr radius[xi]).

That is, if you are measuring the electron is nothing, but eventually it does stop at 1/1836 away from the atom's radius of the measure. You do not get 100% repulsion in the measurement. It repulses 99.98%, then hits the same limit at the measurement atom. The tiny electron 'mass' measurement is the reflection of the measuring devise, not tiny 'mass' of the electron.

Challenge: Why does 1s Shells keep getting closer (smaller radius) when elements are increasing the magnetic field? Shouldn't the 1s Shell get further away?

A great question. However, that misses that two components that change at higher elements. Let's state the challenge as two factors in a typical element (006-C Carbon) versus one element with one more proton (007-N Nitrogen):

Change in Charge, Magnetics and Shell-1

Element	Protons – Charge Attraction	Magnetic Particles (protons + electrons)	1s2 Shell radius (using Ionization as inverse)
006-C Carbon	6	-12	1/1086.454 = 0.0009204
007-N Nitrogen	7	-14	1/1402.328 = 0.0009594
Increase	+1 (+14%)	+1 (+14%)	-0.0021 (-22%)

The Proton Charge attraction increases, and the overall magnetic field increases by the same ratio. If they are the directly related, then the 1s2 Shell should stay in exactly the same place. Right?

But the shell distance (inverse of the ionization energy) actually decreases. This would seem to get a deal breaker.

While the magnetic field gets stronger, the 1s sit at north-south which are the low areas of the field. As a result, there is not much if any of the increase because magnetic field at north-south in the hole.

As a result, the attraction (more protons) increases, but the magnetic field – at the magnetic pole – does not change.

Change in Charge, Magnetics and Shell-1

Element	Protons – Charge Attraction	Magnetic Particle Strength (only N-S)	Ratio	1s2 Shell radius (using Ionization as inverse)
006-C Carbon	6	-12	-2.00	1/1086.454 = 0.0009204
007-N Nitrogen	7	-12	-1.71	1/1402.328 = 0.0009594
Increase	+1 (+14%)		(-14%)	-0.0021 (-22%)

Now, the solution has the correct sign, but not quite there.

Outer Shells Push In More

There are also outer shell electrons which push the 1s2 closer, but ½ are on the other side of the atom so don't count.

The average as discussed moves up, but the impact of other electrons in Shell-2 actually pushes the Shell-1 closers. The average can still increase, but the first layer can go the opposite direction.

Change in Charge, Magnetics and Shell-1

Element	Protons – Charge Attraction	Magnetic Particle Strength	Ratio	1s2 Shell radius (using Ionization as inverse)
006-C Carbon	6	-12	-2.00	1/1086.454 = 0.0009204
007-N Nitrogen	7	-12	-1.71	1/1402.328 = 0.0009594
Increase			(-14.3%)	-0.0021 (-22.5%)
Pressure more Electrons in outer shell	(-14%)/2		(- 7.2%)	
Calculated change to 1s2 radius			(-21.5%)	(-22.5%)

That gets us close enough. It certain shows a reason why the s-shells would decrease in radius as the total protons and mass of the elements increases.

Fundamental Question: What Happens with Temperature? I see magnetics decrease, but gravity does not?

This topic goes back to Book 1 – Nucleus Magnetic Chain-Ring.

Temperature is just increases in movement. Specifically, the movement can be of two types:

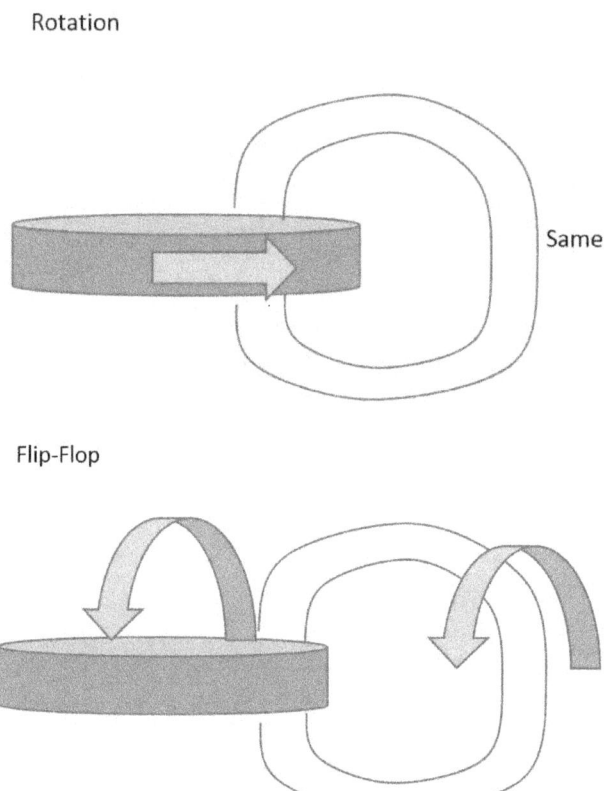

Or more likely a combination of those.

What happens is that with more flip-flop, the magnetic field keeps moving. Over the periodic cycle of the flip-flop, the strong goes weak, and the weak gets strong, then repeat the cycle. That means the difference in magnetic strength keeps changing. Over many molecules with different flip-flops eventually at high temperatures, the north-south gets normalized. That is, what made one direction north gets replaced, but 10 locations as north 10% of the time.

For the impact on other objects, once you get to higher temperatures, then the object is both pulled and then not pulled, or even pushed. But at different times other atoms are pushed, then not pushed. The forces cancel each other.

Again, our measurement of the magnet is an integral over time. In fact, the total magnetic force has not changed, but the differential which is what must reach a threshold to 'move' something keeps going down.

We live in a huge magnetic field. The Earth's. However, that does not move us all the time. That is, because that field is about the same strength a mile north of us and a mile south of us. We are not in a differential, but in a steady location.

Fundamental Question: What Explains why Mass Changes in Bonding?

Wikipedia says:

"Classically a bound system is at a lower energy level than its unbound constituents, and its mass must be less than the total mass of its unbound constituents."

What changes is R_{ES}, not N.

When you add another atom and share the electron between them, that structure does not have the protons all in the middle. That changes the R_{ES} calculation. The space between new combo-nucleus is a dead zone, and that is exactly where the bonding electrons sit.

The total # of N nucleus particles does not change, but number of those particles that contribute and how they contribute to the gravity calculation do change. The bonded electrons stop part of their contributing to the gravity formula.

To me, N is the source of mass, not N/R_{ES}^3 which is the observed mass.

The extra nucleus in a different centering point changes the average distance of the set of electrons. This factor is smaller, but offsets the bonding lack of contribution.

Changing the average distances creates a slight change in the Charge/distance-square electron-vs-nucleus differential. That changes the force projected to other atoms – the 'mass.'

As such, the differential is slightly different, and we can measure the change in 'mass' which is a change in the average electron distance of the combined molecule versus the atoms operating autonomously.

Bonded Electrons do not contribute to the external Atom-gravity 'mass'

A picture explains. In an atom, the electrons must be all pushed out relative to the nucleus. That gives a consistent ratio which measures as a steady mass. However, in a molecule, that is bonded atoms, the nucleus sits in different places. As a result, those electrons that are between the nucleus1 and nucleus2 do not contribute to gravity. In fact, they have the reversing effect, the nucleus are outside those electrons so that reduces the atom-gravity.

Bonding Electron Not Contributing to Gravity / Mass

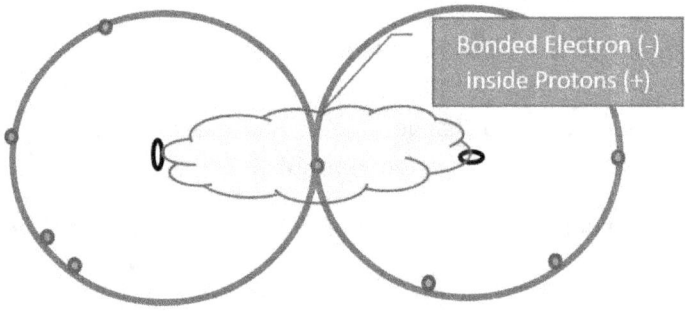

The electron in the bonding position does not contribute to charge-differential (atom-gravity 'mass'). In fact, it decreases the calculation overall. Instead of electrons outside protons consistently, this event is electrons inside protons.

Bonding – As If Electron Removed From Mass

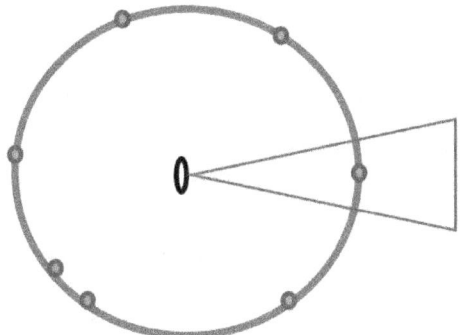

But there are 'reflections' that mitigate that. One great set of mathematics described something similar as the Fresnel zone. In my real job, I have built cellular telecom systems and more specifically microwave towers to connect those towers. We needed to check Fresnel for each tower. Microwaves fail even if the line of sight is clear because of the other 'Fresnel zones.' Between two towers in a narrow window (like the electron-bond cloud), in increments of 1/n wavelengths, that Fresnel gets calculated, if there are any objects in that red cloud, the signal gets messed up. That calculation can get messy, but generally, all you really need to know is do not let trees even in the 'curved space' between microwave towers. A tree, wall or parked truck even not directly between can creates reflections that change the base microwave operation.

The slightly different logic in the details applies to atom-gravity 'mass', but it has similar complexity. While the electron does not contribute to charge-differential (atom-gravity 'mass'), there are 'reflections' that mitigate that.

One would expect that the 'mass' lost in bonding would get reduced by that reversal of:

Electrons bonding / Total # of Electrons

However, experimental evidence shows the change is not that dramatic.

Decrease offset in that Electron still contributes to push out other electrons

It is one step more complex. While that electrons does not create a charge-differential, it does contribute to keep the other electrons out in their orbit which means that the average radius does not decrease.

If you take out that electron, then the other electrons would have been closer in. That electron, even in the bond, still contributes to the electron-shell radius (distance for 1/distance-squared).

Without that bonding electron, the other electrons would have position closer together. Without that electron (red triangle) the other electrons (purple) would not have the electron-electron repulsion (orange arrow) from that missing direction. As a result, the bonding electron still does some contribution to the electron-shell radius.

Bond Electrons Do Add Part of Mass that Pushes Other Electrons Outside Proton-Proton Vortex

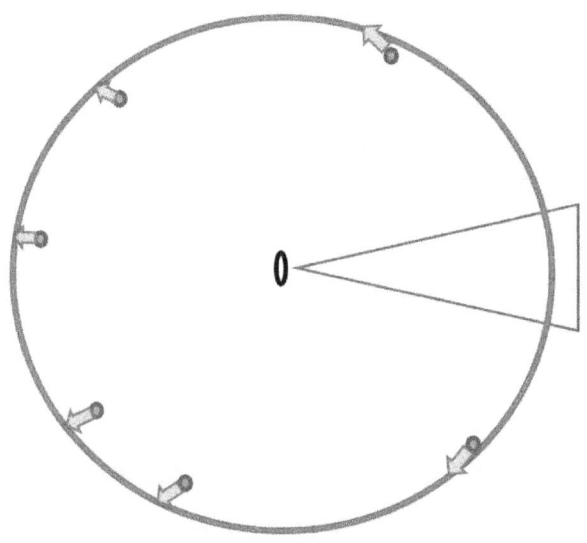

The electrons-in-bonds impact on shell distance can get describe as:

	Electron-Electron Repulsion	Electron-Electron Repulsion	Change to atom-gravity
In Atom	100%	100%	
When Bonded	100%	(n-1)/n	Pro rata

Decrease offset by increases in distance to electron in other atom in the bond

In addition, the bond distance also now goes in part to both nucleus locations. That makes distance (and thereby atom-gravity) increase.

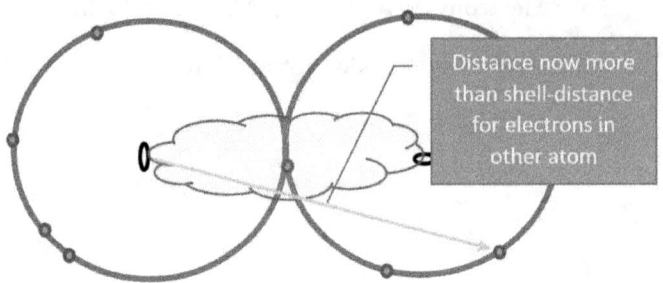

Must also realize location is not fixed, but a quantum path

Further, each electron floats, so, in its quantum path, that electrons actually is outside that zone some percentage of the time and does contribute in those periods.

Calculations done with a perfect circular orbit or exact electron location yield a result up to a factor of 0.6 too small. However, in quantum-path, that electron predicts to be closer and farther away. However, again the 1-distance-squared makes the times when quantum-walk closer have huge impact.

Conclusion: Too much to calculate here

So, I will report in a few years and update this book when all those elements get calculated. I have not done those.

However, in generally, the atom-gravity ('mass') of the combined molecule must go down when compared to the individual atoms. I expect the mass reduction is a Frenzel zone calculation. Those electrons shared contribute less than standard to the charge-differential calculation that is the basic of atom-gravity ('mass').

Fundamental Question: What Explains why Mass Changes in various Einstein equations?

This work is still in progress. I hope that scientists, including me if I have time, will solve the Einstein Relativity equations replacing time-space variability with electron-shell-distance and time-integrals. That is, when we change 'mass' or 'time' that the same results can get achieved by instead changing, for a physical reason, R_{ES}. R_{ES} changes will impact the final result in exactly the same math way. The excellent math is still an amazing feet and discovery. The trick is splitting 'mass' by the new item will apply all the same tricks to a physical factor.

My proposal is that whenever the shortcut of making mass, time or space variable occurs, that same calculation can occur by making the R_{ES} charge-differential via changes in the electron distances, magnetic field, and quantum path complexities do the same thing. All the extra dimensions of various string-theory and such resolve to known dimensions.

$$mass = \frac{G_A N_1}{\sqrt{G_N}\left(\frac{4}{3}\pi R_{ES}\right)^3} = \frac{M(Z,N)_1}{\left(\frac{4}{3}\pi R_{ES}\right)^3}$$

This does not mean that Einstein's photoelectric effect is wrong. That one is amazing. The proof that the universe operates a specific 'quantum of energy' remains a bedrock standard.

It just means that the Special Theory of Relativity and the General Theory of Relativity need an update for this split definition of mass and gravity. There will be formulas that address at least the Special examples that have factors in basic space and electromagnetic forces substituting for the curvature elements.

That the stretching effect has a physical structure and field energy source to replace the mathematic contortions in relativity equations.

For example,

Atoms approaching near speed of light build towards infinite 'mass'

Here is a great example, as atoms approach the speed of light, their 'mass goes up'. Great physicist like Lorentz documented the calculations including 'mass increase'. Experimental evidences has tested and validated 'mass' increase.

However, I propose a physical reasoning that would increase mass. When moving at the speed of light, the electron cannot keep up, so more and more electrons misses lots of the outer (reduced reflection) positions in the electron-shell. What is left is an tighter shell (orbit) which creates higher net-electrostatic. *As R_{ES} decreases down, $\frac{1}{R_{ES}}$ increase – 'observed mass' goes up.*

The Gauss field stops being homogeneous because the half moving forward is very different than the half of the orbit when the electrons move back.

Around the atom, the electrons sit in little valleys of field-energy. But the underlying atom is moving and rotating and flip-flopping so the electrons follow a quantum-path. That path includes a constant pull to the nucleus (orange arrow), but along a moving, even changing potential (blue line).

Electrons Sit a Different Distances by Field Strength

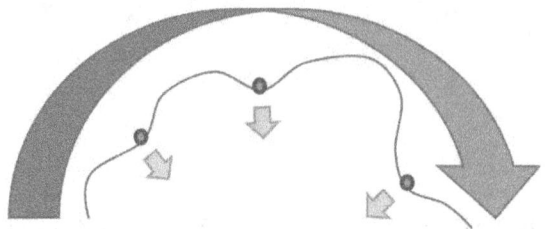

When the entire system is hardly moving, the electron sits at the best spot in the electron-shell. The hills and valleys are the magnetic-repulsion potential around the sphere (as 2D map in paper for better illustration). In the below visual, the election sits in the low portion of the field.

Distance Lowest When Atom Not Moving

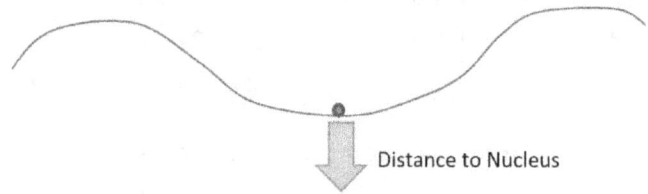

As the whole atom structure moves, the electron cannot keep up so it moves up the potential until it can catch up. This happens a lot when the whole atom is moving at a percentage of the speed of light. At no movement, the electrons settle into the lowest distance (most proton pull) to the nucleus. However, when the atom moves, the electrons cannot keep up, so the 'roll up the back (opposite the movement)' the potential-curve (which also drives the nucleus distance). *A ball would roll back and up as the wavy carpet moves forward.*

More Distance (Electron to Nucleus) When Energy Moving Atom

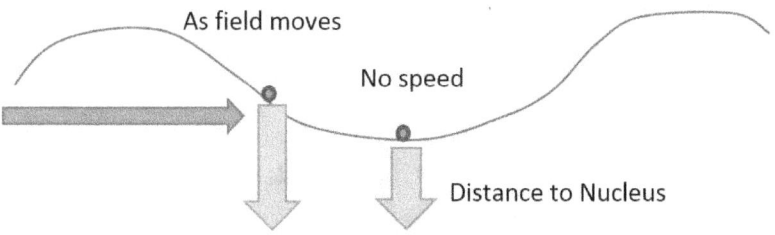

That extra distance to the nucleus is one of the key elements of atom-gravity ('mass') calculations. As electron-shell distance increases, then the 'mass' increase without change the physical attributes.

However, that explanation is not a change to the underlying Charge of the atom(s) or its basic particles. That are still P # of protons, N # of neutrons, and E # of electrons (E matching the # P protons in stable atoms).

Of course, this phenomenon falls apart when the electron cannot keep up. The electron has another path and that is to leave this atom and eventually catch another. So, there is a physical limits and alternatives beyond a basic range that Einstein and Lorentz did not address. The science fiction application can never occur.

Part 2 – Electron Speed in Orbit More than Speed of Whole Atom

If an atom rotates and move both at some percentage of speed of light, the electrons definitely cannot keep up, and stay in the low-potential distance to the nucleus.

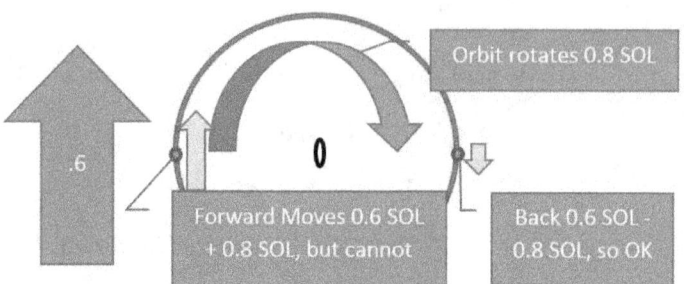

During the electron orbit portion in the direction of the orbit, the speeds combine, and that might run up again the Speed of Light (SOL). As a result, the 'time' of the orbit actually gets stretched. The electron cannot orbit because during that period the system is moving 0.8 and the electron orbiting at 0.6 SOL, which would be 1.4 SOL, but the electron has a physical limit of 1.0 SOL. The 'clock' which operates by the consistent electron orbits starts to slow.

Therefore, the mass increase is not unknown. It can get complicated; I just don't have the time in this book.

Further, the 'time' of electron orbits also increases – which is observed as 'slowing' of measured time.

Fundamental Question: If Gravity is an Integral over Time, then are there Gravity Waves?

Probably, but the likely challenge is that those waves are in the frequencies of the electrons shells, so there is not one gravity specific wavelength that NASA can look for in the universe.

Further, temperature is a factor. This gets beyond the scope of this document.

Can Gravity Waves Synch?

Probably, that would be interesting.

For the Daring

The detailed calculation is a little different, because of the averaging over time and over spheres (integrals), there is a couple of extra 1/2's because electrons on the far side, a 3/4π and 2 because gravity has two (2) atoms, and another 2 because the charge moves from +1 to -1, not +1 to zero. Further, it is a challenge to determine whether the electrons move in synch (d) or with electrons (d-electron-shell). The final calculation calculates over the quantum-path area so $4/3\pi R^2$ is also not quite correct; that volume standard applies only if the electrons were in a gravity-orbit, not a quantum-path. Further, there is a stub -3dR and +3dx in the denominator. However, the above gets you through the overall logic.

I welcome some discussions about the integration and the various ½ factors. It was very difficult integral mathematics, and I need to get checked if I got them all correctly.

More formally, the atom-gravity contact is better stated understanding that prior calculations were multiplications, so the root G_A applies.

- Further, the actual calculation gets the proper 4/3π for the volume of a sphere, not a cube/box of just R^3.

- Further, the calculation is based upon two atoms interacting, so we have a further ½ or the square root to apply when you compare G to the new calculation.

- Further, in mathematics the integral of -3 exponent (strength field) generates a further -2 factor. As we move from fields to force, that extra work is required.

- Finally, there is a factor for the electrons on that backside. So, half of the calc is reducing distance-squared, so another factor enters into the calculation.

If you can properly do that force-integral and get all the factors correct before that chapter, you are a master mathematician. My greatest fear is that I will discovered all of this and forgotten one of the factors someplace. That is, I get the logic correct, but my formula is missing ½ someplace. If I did, please forgive me, and praise to any and all those people that can check all of this.

Atom-Gravity

Factor	Factor	Field Strength decreases by	Radius
Nucleus magnetic field vs charge force balance	Gauss	$\dfrac{G_A(Z,N)_1(Z,N)_2}{(4/3\pi)R_{ES}{}^3} = \dfrac{kQ_1Q_2}{2R_{ES}{}^3}$ or as substituted below: $(\dfrac{G_A{}^2 M_1 N_2}{kQ_1Q_2 8/3\pi}) = \dfrac{2R_{ES}{}^3}{2R_{ES}{}^3} = 1$	Nucleus ***
Electrons has closer Electrostatic Charge force	Coulomb	$\displaystyle\int_0^t \int \int (\dfrac{kQ_E}{2(d \pm R_{ES}\cos(\Theta)}$ $-\dfrac{kQ_P}{2d^3})$ $\displaystyle\int_0^t \int_{d-R}^{d+R} \int kQ_E$ $*(\dfrac{1}{2((d-R_{ES})^2 + 2dx)^{\tfrac{3}{2}}} - \dfrac{1}{2d^3})$ $\dfrac{kQ_E}{d^3 \tfrac{8}{3}\pi R_{ES}{}^3}$	Electron Shell *** ****

Factor	Factor	Field Strength decreases by	Radius
		Charge as integral over time over quantum-path electron shell separation	
Distant body recognizes only the 2nd larger element	Product	$\dfrac{G_A N_1 N_2}{kQ_E 8/3\pi R_{ES}{}^3} * \dfrac{kQ_E}{d^2}$ After substituting, strength is $\dfrac{-(G_A N_1)(G_A N_2)}{\frac{4}{3}\pi R_{ES}{}^3 d^3 {}_{+T}} = \dfrac{-2Gm_1 m_2}{d^3}$ Taking the integral, force is $\dfrac{-(G_A N_1)(G_A N_2)}{(-2)\frac{4}{3}\pi R_{ES}{}^3 d^2 {}_{+T}} = \dfrac{Gm_1 m_2}{d^2}$ $$F = \dfrac{(G_A N_1)(G_A N_2)}{\frac{8}{3}\pi R_{ES}{}^3 d^2 + T}$$ or $$F = \dfrac{G_A{}^2 N_1 N_2}{\frac{8}{3}\pi R_{ES}{}^3 d^2 + T}$$	Galaxies away $R_{es}=$ Radius electron shell

**** I use R_{es} versus r because this is the fundamental of electron shell force. It is consistent over all elements in the periodic chart (with known exceptions for Einstein, Lorenz transformations and such).

*** I am only showing the $\dfrac{1}{(R_{ES})^3}$ when there is a tiny tail (T) in the $\dfrac{1}{(d-R)^3}$ for the cross multiple $-dR$ and for the integral $+dx$ segments, but those are a factor $(R)^{-12}$ smaller so in the general presentation they were excluded, but I will combine them in 'T' for 'tiny' and use it only when that precision is required.[xii]

The N = nucleus magnetic particles (protons + neutrons) in the new formula equates to the m = mass in the Newton calculation.

'T' is that tiny calculation from the -dR and +dx factors.

The $\dfrac{G_A}{\frac{4}{3}\pi R_{es}^3}$ = Shell separation field strength in the new formula equates to the Gaussian enclosed electron shell strength.[xiii] Outside the shell, that force net, the electromagnetic Charge offsets all magnetic repulsion. That equality is why the magnetism can substitute for net-charge.

The Charge segment delivers the $\dfrac{1}{d^2}$ the 1/distance-square segment in the Newton calculation, but the kQ_E segments cancels out, so actual charge is not part of the final equation.

It updates Newton without 'mass' in the calculation. The only factors are known constants: N = actual nucleus particles in atoms and R = average radius of electron shells based upon the nucleus repulsion which is really a constant field strength placing electrons in a shell radius*.

* with notable adjustments Einstein, Lorentz, etc.

Conclusion: Gravity is excess Charge (1/distance-squared) applied based upon separation (linear) of charge elements via a push from the magnetic field of nucleus particles (protons + neutrons).

Future Work for the Scientific Community:

I expect that G_a will get replaced by a Magnetic Force Constant in further studies. Maybe, I shouldn't use the letter G, but M. M = Magnetic Force Constant. However, for this point of introduction to solve gravity, the G makes the relationship clear and the direct connection to replace the old G. G_a is the Gravity from an atom, which I believe is really its magnetic field.

The scientific community will need to investigate, test, and validate the additional postulate that G = M. I have not completed and documented that in time for this book.

Of course, there is a further postulate that the Magnetic Force Constant is directly calculated from k, the Charge Force Constant. Again, that further postulate is not completed and documented in time for this book. That should be fun, if you are an obsessed science junkie. That is the next book in this series.

For the Physics and Chemistry Community:

As a result, the properly reduced G_A and G_A^2 has applications within electron-shell calculations, chemical bonding, and probably even the core magnetism.

For use with interacting atoms, the G_A^2 will apply. For use within just one atom, the G_A will apply.

Like Planck and reduced-Planck ($\hbar$), which is the square root of Planck, this new constant in both forms, itself and the square, will appear in many future formulas and calculations.

Background from Prior Books

To understand the electron shell segment of the formula, it probably helps to know the core concepts from the earlier books in the Simple Words™ to Understand . . . Chemistry, Elements, and Bonds series.

Book #1

Every nucleus has protons with positive (+) charges.

Every nucleus creates a magnetic field.

It is that magnetic field that make the nucleus stay together despite that protons strongly repel each other. A nucleus is really just a long magnet of various structures, orientations, and shapes. When in a continuous chain or ring that separate protons from neutrons, the magnetic strength remains higher than the proton-proton repulsion.

Patent pending 15256865 which add magnetism to the structure of nucleus chemistry sets by: a) magnetic links that create a chain or ring (1-4) that holds the nucleus together; and b) overall magnetism (5) of nucleus-ring structure to get applied elsewhere.

Figure 12

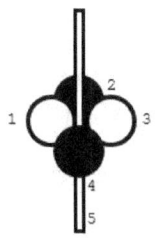

Book #2

Electrons have negative (-) charge which attracts the electrons to the proton's positive (+) charge in the nucleus.

Electrons (-) are repelled by magnetic fields – north or south.

Electrons don't fall to the nucleus because they are 'held' out in 'orbit' by that electron's repulsion to the nucleus-magnetic-field.

At the shell distance, that repulsion balances the negative-charged electron (-) to the positive-charge (+) proton attraction. So, electrons stay away in 'orbit' at a consistent distance.

While Charge is always stronger than magnetism:

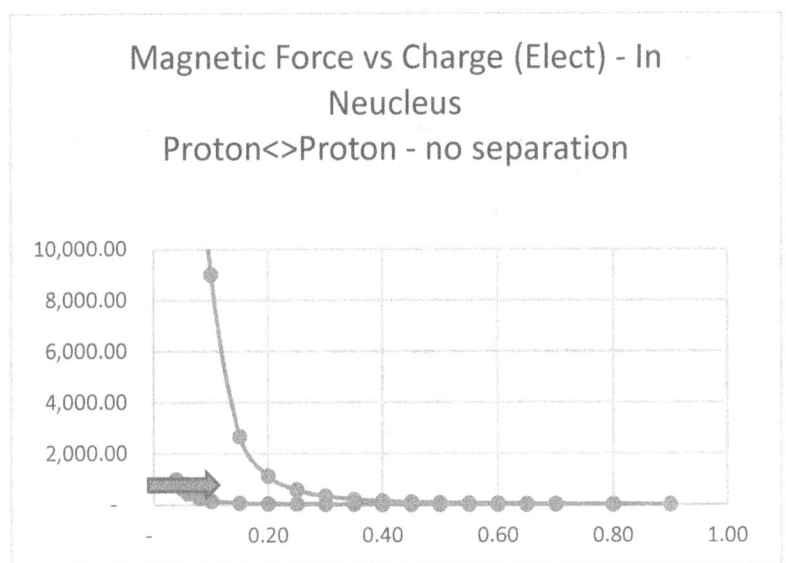

Because a magnetic field does not start decreasing until the physical chain is broken, there is proton-neutron-proton configurations that allow a nucleus to stay bonded together.

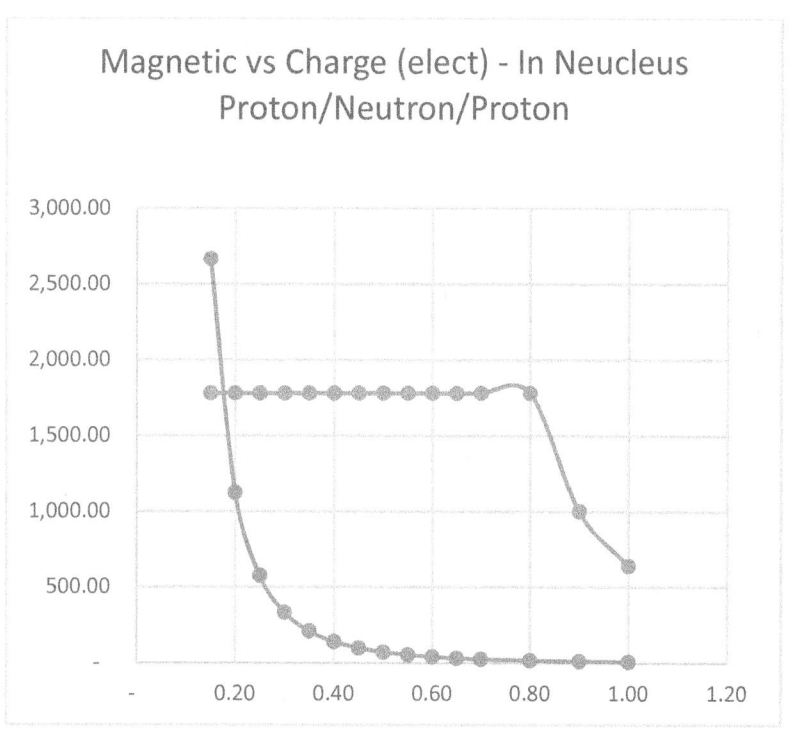

Book #3

Thereby, the electrons sit in magnetic scrunched cube 'shells.'

Patent pending 15245326 chemistry-set with bond-angles and magnetic orientation shows how 018-Ar Argon electrons settle (in low energy systems).

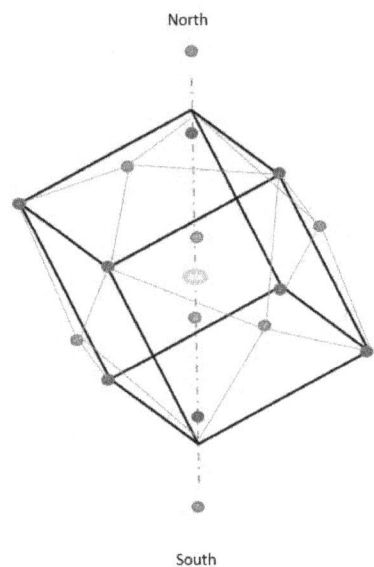

That electrons sit is shells that balance the charge attraction (electrons to protons), with the magnetic repulsion (electrons to the overall nucleus magnetic field becomes the key fact for the actual gravity formula in Chapter 5.

Revised Gravity Constant Derivation based upon Charge and Electron Field

With all due respect to Newton and later scientists, the calculation of gravity and mass can get derived as part of the small leftover balance of charge and magnetism (called electromagnetism) of the atoms involved.

Of course, everyone know that gravity is directly related to mass. What everyone does not understand is the mass of an atom, and its component particles flow directly from the electromagnetic properties of each atom, specifically the electron-shell. Effectively, there are four basic interactions to calculate to understand atom-gravity. These are the interaction for the two (2) element (proton and neutron) between two (2) different molecules at a distance:

	Direction	Calculation
Nucleus1 <> Nucleus2	Repulsion + / +	Nucleus to Nucleus distance
Electron1 <> Nucleus2	Attraction - / +	Integral over electron location - complex
Electron2 <> Nucleus1	Attraction - / +	Integral over electron location - complex
Electron1 <> Electron2	Repulsion - / -	Integral over electron location - complex

Math Formula

Nucleus1 <> Nucleus2 $\qquad \iiint_0^t \dfrac{k\, Q_1\, Q_2}{(d)^2}$

Electron1 <> Nucleus2 $\qquad \iiint_0^t \dfrac{k\, Q_1\, Q_2}{(d + (r + \cos(x))^2}$

Electron2 <> Nucleus1 $\qquad \iiint_0^t \dfrac{k\, Q_1\, Q_2}{(d + (r + \cos(x))^2}$

Electron1 <> Electron2 $\qquad \iiint_0^t \dfrac{k\, Q_1\, Q_2}{(d + (r + \cos(x))^2}$

From here forward, the math side of the calculation will get demoted to the end notes. If you really want that painful process, then please take a look. In fact, for the adventurous, the last chapter will provide a series of experience still required to 'prove' the theorem to the level enough for Nobel Prize nomination. I expect that process will take a decade.

Explaining Why Electromagnetism was Missed as the Basis of Gravity

Two main reasons exist why electromagnetic charge of each atom was not considered as the basis of gravity:

1) Electromagnetic charges offset over the atom as a complete system. This means that calculations always netted to zero when the atom is taken as a whole.

2) Experimental evidence shows a measurement called mass (protons + neutrons) fit a general calculation for the calculation of gravity, but mass of atoms does not follow the growth of electromagnetic charge (protons only).

For centuries, scientists have looked at the basic structure of an atom as Electromagnetic charge **neutral**. There are an equal number of protons (+ positive charges) and neutrons (- negative charges). As such, the calculation was always that the positive protons equal the negative electrons which nets to zero. Therefore, the thinking is that Charge is not involved in gravity.

A zero net charge seems to give nothing to use in calculations. 5th grade taught us that anything times zero is zero.

Further, mass which has been proven as an excellent measure for calculations involving gravity is not based upon charge. You can see this in the table from the periodic chart below:

Element	Charge	Atomic Mass	Ratio
001-H Hydrogen	1	1	1
002-He Helium	2	4	2
003-Li Lithium	3	6	2
004-Be	4	9	2.25
005-	5	11	2.2
006-C Carbon	6	12	2
026-Fe Iron	26	59	2.1

Anyone can see the huge jump in Mass versus Charge from Hydrogen to Helium. The Charge goes from 1 to 2, but the mass goes from 1 to 4. This variance was extremely well known since the observations of gravity used the Sun and planets, and the Sun is almost entirely made of H Hydrogen, but the planets are made of elements with higher elements.

The calculations worked only using the Mass column, not the Charge column. Therefore, any calculation with Charge would seem from the start as a total waste of time.

Gravity is Multiple Particles Forces from Different Atoms Combined

The gravitational force is the integral over any time period that is much greater than (multiples of) the frequency of the orbit of the electron set around the nucleus.

That integral is taken over the charge-force calculation, Coulomb's Law for the various elements:

Atom1	Atom2	Distance	Net Force
Protons in Nucleus	Protons in Nucleus	d	Repulsion
Protons in Nucleus	Electrons in Shell-structure	d+/-Integral-electron-orbits	Attraction
Electrons in Shell-structure	Protons in Nucleus	d+/-Integral-electron-orbits	Attraction
Electrons in Shell-structure	Electrons in Shell-structure	d OR d+/-Integral-electron-orbits	Repulsion

- Proton<>Proton repulsion at the basic distance;

- Proton<>Electron attraction at integral of the distances as the electron's move both closer and further away – twice as many as this is a cross-intersection – Atom 1 to 2, and 2 to 1.

- Electron<>Electron repulsion on the complex combinations of both electrons sets orbiting

$e = mc^2$ Using Nucleus Particles instead of Mass

However, looking at the first section makes gravity based upon electromagnetic charge, not the traditional word - mass. Yet, Newton used mass as the basis for his gravity calculations, not electromagnetic Charge. Further, mass is key of so many well-known formulas, like Newton's gravity or Einstein's energy limitations:

And we have Einstein's

$$e = mc^2$$

And we have Newton's

$$F = \frac{G_N m_1 m_2}{d^2} = \frac{\sqrt{G_N m_1}}{d} * \frac{\sqrt{G_N m_2}}{d}$$

and the revision

$$F = \frac{\sqrt{G_N m_1}\sqrt{G_N m_2}}{d^2} = \frac{(G_A N_1)(G_A N_2)}{\frac{8}{3}\pi R_{ES}^3 d^2 + T}$$

Ignoring T, the tiny tail, we can solve for:

$$\frac{\sqrt{G_N}m_1}{d} * \frac{\sqrt{G_N}m_2}{d} = \frac{(G_A N_1)}{(\frac{8}{3}\pi R_{ES})^{3/2}d} * \frac{(G_A N_2)}{(\frac{8}{3}\pi R_{ES})^{3/2}d}$$

But, the two factors should be in ratio, so let's solve for only one side to the one on corresponding:

$$\frac{\sqrt{G_N}m_1}{d} = \frac{(G_A N_1)}{(\frac{8}{3}\pi R_{ES})^{3/2}d}$$

$$m_1 = \frac{(G_A N_1)}{(\frac{8}{3}\pi R_{ES})^{3/2}\sqrt{G_N}}$$

Then this Einstein formula can get replaced with the below:

$$e = \frac{(G_A N_1)c^2}{(\frac{8}{3}\pi R_{ES})^{3/2}\sqrt{G_N}} = \frac{M_{NP}N_1 c^2}{(\frac{8}{3}\pi R_{ES})^{3/2}}$$

Now, we have three factors that are known instead of mass. Mass is not part of the equation.

The number of nucleus particles, **N**, does not change. At the end of the interactions causing space-time calculations, we still have the same element as each atom. We know the ending atom is exactly the same as the initial atom in number of protons, neutrons, and electrons.

G_A and G_A^2 do not change as this is a fundamental property of electromagnetism, and thereby gravity. However, this is not certain, as some interactions may change the G_A as it create M magnetism for electron repulsion.

In all likelihood, R_{ES} changes. That is a physical distance caused by real events. While stable under most conditions, most of the examples of Einstein and Lorentz can change this R_{ES}.

However, there really is no inconsistency. In fact, the first part above is a resolution to that link (between Electromagnetic Charge and Mass) as mass is the calculation of that net electromagnetic force from a large set of atoms.

What happens to link the charge to the mass is quite simple. The distance of the electron shell is pushed out by the total magnetic force of the nucleus (this is the new concept) creates a 1/distance squared distance. Those two are really the same. Both are calculated by the same combination of protons and neutrons. The distance of the electron shell is driven by the magnetic force from the nucleus. That magnetic force is the combination of protons and neutron. Yet, the combination of protons and neutrons is also the core definition of the mass of an atom. [A later chapter will discuss the 'mass' of electron]

- Protons + Neutrons => Magnetic Force of Nucleus => Electron-Shell Distance (average radius) => Charge differential force of 'gravity'

- Protons + Neutrons => Mass of Nucleus and Atom (understanding that most people think of the electron as near zero)

Impact of Direct Calculation of Gravity from Electrodynamics on String Theory and Such

> A Kiss
>
>
>
> Gravity and some of the other forces, and subatomic particles in current theorems, like 'string theory', I compare to a kiss. A kiss is a real thing. We count how long it lasts. We can take pictures of it. We can obsess about the best ones. We want more and more.
>
> However, a kiss is not a person. Gravity is not a particle; it is a special event or net-combination.
>
> When the scientist get all excited that they created a new particle that last 11 nanoseconds, I think of a kiss. Great that they could create that. Horray!
>
> The same applies with ocean waves. That is both particle and wave. Let's not confuse the particles of water with the amazing even called a wave. Don't think that because water can form a wave that we have wave-particle duality.
>
> Let me play with photons, neutrinos, electrons and other particles that last more than a few nanoseconds. Give me the person causing the kisses and I will have kisses that lasts forever. Ah!

The challenge today is that many other people have calculate using a system of 'string theory' or others with invisible dimensions and other magic, but unknown particles, like 'Higgs bosons', which create this calculation of various observed forces as from one of these separate 'dimensions' in the calc. The goal here is make it clear that existing standard factors address this, that many of these 'string theory' extra dimensions are simply a complex combination of base functions.

String theory has seven (7) or eleven (11) or many other versions.

Those flow beyond the basic four:

Geometry – Cartesian (or Spherical)

    Length (Longitude)

    Width (Latitude)

    Height (Azimuth)

Energy

    Electromagnetism fields and forces

The proposition here is that various, maybe even all, of those additional 'dimensions' are derived from electromagnetics and geometry of the basic four. Between this gravity theorem herein; the nucleus-strong-force theorem of Book 1; and the electron-shell theorem of my other book, each of the 'string theory' extra dimensions get replaced by functions strictly from the basic geometry of the base particles and their energy known factors.

- Gravity is not some special new dimension.

- Spin is not some special new dimension.

- Strong force is not some special particles or new dimension.

- Color is a real thing, but not some new special dimension.

These become further evidence that string theory may solve certain calculations, but the 'special dimensions' are not some actual change to our understanding of the universe. Extra dimensions theorems are very useful because they solve problems quickly. However, those are not the ultimate; they are corollaries.

Calculating the Current point-distance Gravity into a Net-Charge Surface Integral Not via Field Strength Shortcuts

We will start with the basic charge calculation:

$$F = k_e \frac{Q_1 Q_2}{d^2}$$

That is a point calculation, so going with charge box of 1x1, then the integral is:

$$F = \int_{x=d-1\,Planck}^{x=d} -2k_e \frac{Q_1 Q_2}{x^3}$$

Yet, the Arno-Gravity postulate actually changes this point-force into separate the charge into the protons and the electrons as:

+ 2 Electrons-Proton Attraction

− Proton-Proton Repulsion

- Electron-Electron Repulsion

But I will assume rotation of electrons moves in synch to simplify this segment as the same as Proton-Proton Repulsion.

+ 2 Electrons-Proton Attractions

− 2 Like-Kind Repulsions

Further, because the electrons sit at the shell distance, this is really over-a-surface calculation, not a volume integral. (I am ignoring the quantum-path factor for the moment. Yes, r is not a gravity-orbit. We will attempt to address that complexity later.)

$$F = 2\int \int -2k_e \frac{Q_1 Q_2}{x^3} dS - 2 \int_{x=d-1 \, Planck}^{x=d} -2k_e \frac{Q_1 Q_2}{x^3}$$

We are trying to determine the surface integral for the distance from a far point to an electron sitting on the surface of the electron-shell, and that electron-shell is a sphere, so this becomes a direct calculation based upon these x,y,z determined by the following drawing.

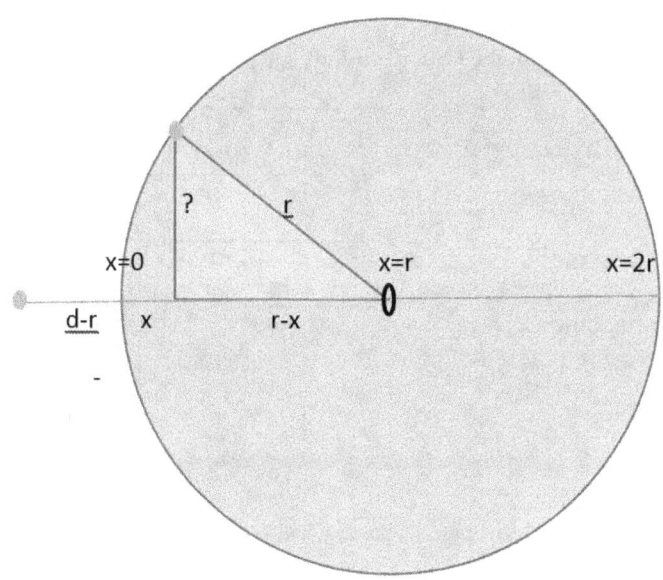

$$F = \int_{x=d-r}^{x=d+r} \int_{y=-?}^{y=+?} \int_{z=-?}^{z=+?} \frac{-2k_e Q_1 Q_2}{distance^3}$$

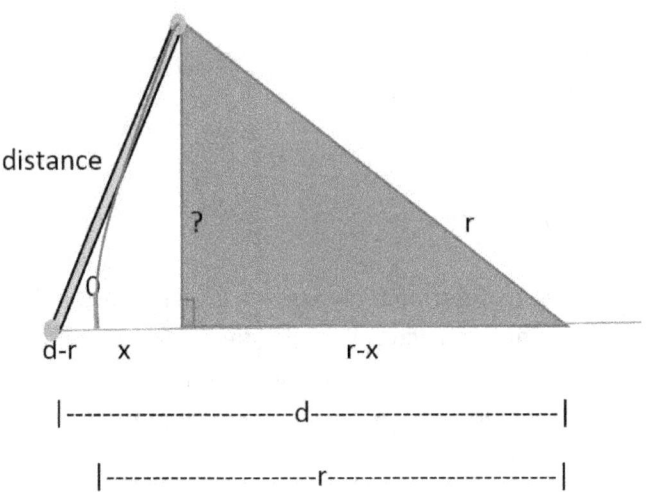

Solving for the height (? In drawing) is the square root of:

$$? = \sqrt{r^2 - (r-x)^2}$$

which becomes $\qquad ? = \sqrt{r^2 - (r^2 - 2rx + x^2)}$

which becomes $\qquad ? = \sqrt{r^2 - r^2 + 2rx - x^2}$

which becomes $\qquad ? = \sqrt{+2rx - x^2}$

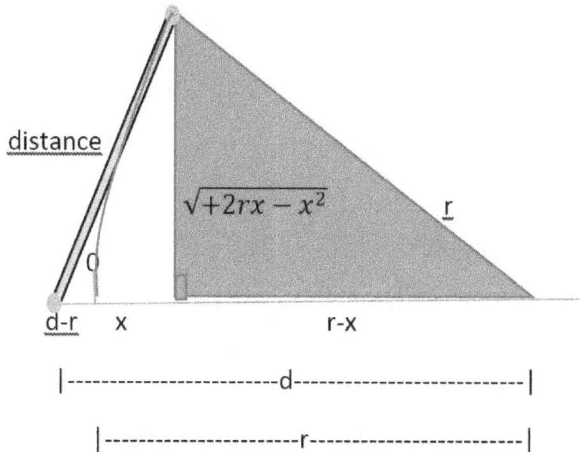

We can use the Pythagorean Theorem for the d=distance, and since we are 1/distance-squared the square root at the end gets squared, so the result is cleaner.

The base at a right angle as designed:

$$\text{Base} = (d - r + x)^2$$

which becomes $+d^2 + r^2 - 2dr + 2dx - 2rx + x^2$

So, the distance is the sum of the squares, but we won't need the square root since the 1/distance squared just redoes that step later anyway.

$$d^2 + r^2 - 2dr + 2dx - 2rx + x^2$$

The applying the Pythagorean Theorem to the left triangle, we get:

$$\sqrt{+a^2 + b^2} = \sqrt{+2rx - x^2 + b^2 + r^2 - 2dr + 2dx - 2rx + x^2}$$

But the two yellow items cancel out to the same in the other leg (as purple):

$$\sqrt{+2rx - x^2 + d^2 + r^2 - 2dr + 2dx - 2rx + x^2}$$

Which then simplifies to:

$$\sqrt{+d^2 + r^2 - 2dr + 2dx}$$

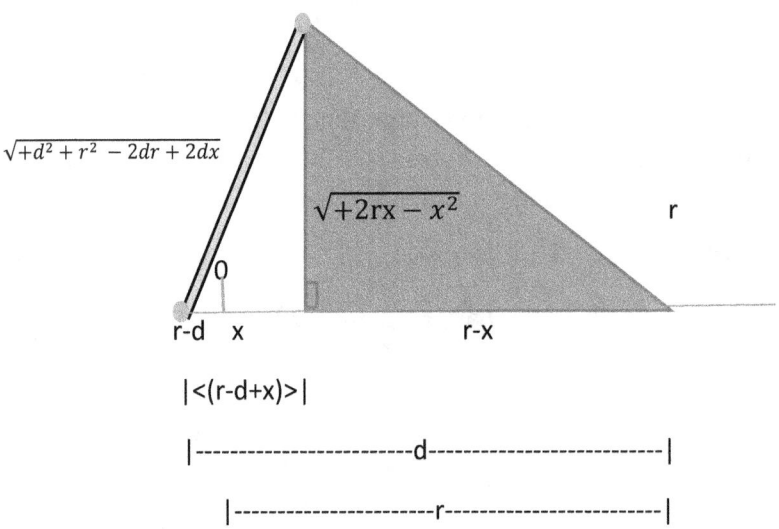

Now that we have the real distance, the charge repulsion force is . Note that at this point, we have moved a) to the entire system with the distant object, and b) using the Force integral, not the field integral only for the electron shell force assuming spherical average distribution.

Given d>r>x, d,r,x all > 0, and $x^2 + y^2 + z^2 = r^2$:

$$\int_{x=d-r}^{x=d+r} \int_{y=-\sqrt{+2rx-x^2}}^{y=+\sqrt{+2rx-x^2}} \int_{z=-\sqrt{+2rx-x^2-y^2}}^{z=+\sqrt{+2rx-x^2-y^2}} \frac{k_e Q_1 Q_2}{\left(\sqrt{d^2+r^2-2dr+2dx}\right)^2}$$

By using the force, we get the square of a square-root which cancels:

Given d>r>x, d,r,x all > 0, and $x^2 + y^2 + z^2 = r^2$:

$$\int_{x=d-r}^{x=d+r} \int_{y=-\sqrt{+2rx-x^2}}^{y=+\sqrt{+2rx-x^2}} \int_{z=-\sqrt{+2rx-x^2-y^2}}^{z=+\sqrt{+2rx-x^2-y^2}} \frac{k_e Q_1 Q_2}{(d^2+r^2-2dr+2dx)}$$

Given d>r>x, d,r,x all > 0, and $x^2 + y^2 + z^2 = r^2$ as the attraction and repulsion:

$$\left(\frac{k_e Q_1 Q_2}{\left(\sqrt{(d-r)^2-2dx}\right)^2}\right) - \left(\frac{k_e Q_1 Q_2}{\left(\sqrt{(d-0)^2-2d0}\right)^2}\right)$$

Now, we can use the distance of the electron attraction versus the proton repulsion.

Given d>r>x, d,r,x all > 0, and $x^2 + y^2 + z^2 = r^2$:

$$k_e Q_1 Q_2 \left(\frac{1}{(d^2+r^2-2dr+2dx)} - \frac{1}{d^2}\right)$$

$$k_e Q_1 Q_2 \left(\frac{d^2 - (d^2+r^2-2dr+2dx)}{d^2((d-r)^2+2dx)}\right)$$

$$k_e Q_1 Q_2 \left(\frac{d^2 - d^2 - r^2 + 2dr - 2dx}{d^2((d-r)^2 + 2dx)} \right)$$

$$k_e Q_1 Q_2 \left(\frac{-r^2 + 2dr - 2dx}{d^2((d-r)^2 + 2dx)} \right)$$

But this must get averaged over the x for the sphere of electron-shell positions.

As the field strength, the calculation is impossibly more complex. That is because the square root is taken to the third power, which gets beyond any easy simplification:

Given d>r>x, d,r,x all > 0, and $x^2 + y^2 + z^2 = r^2$:

$$\int_{x=d-r}^{x=d+r} \int_{y=-\sqrt{+2rx-x^2}}^{y=+\sqrt{+2rx-x^2}} \int_{z=-\sqrt{+2rx-x^2}}^{z=+\sqrt{+2rx-x^2}} \frac{-2k_e Q_1 Q_2}{\left(\sqrt{d^2 + r^2 - 2dr + 2dx}\right)^3}$$

Or in this form:

Given d>r>x, d,r,x all > 0, and $x^2 + y^2 + z^2 = r^2$:

$$\int_{x=d-r}^{x=d+r} \int_{y=-\sqrt{+2rx-x^2}}^{y=+\sqrt{+2rx-x^2}} \int_{z=-\sqrt{+2rx-x^2}}^{z=+\sqrt{+2rx-x^2}} \frac{-2k_e Q_1 Q_2}{\left(\sqrt{(d-r)^2 + 2dx}\right)^3}$$

Putting this back in with two (2) attractions (electron-to-distant-proton and vice-a-versa) and two (2) repulsions (proton-proton and electron-electron):

Given d>r>x, d,r,x all > 0, and $x^2 + y^2 + z^2 = r^2$:

$$\int \int \int -4k_e Q_1 Q_2 \left(\frac{1}{((d-r)^2 + 2dx)^{\frac{3}{2}}} - \frac{1}{d^3} \right)$$

Finally, we can complete the formula with the d = distance in the volume integral formula:

Given d>r>x, d,r,x all > 0, and $x^2 + y^2 + z^2 = r^2$:

$$F =$$

$$\int_{x=d-r}^{x=d+r} \int_{y=-\sqrt{+2rx-x^2}}^{y=+\sqrt{+2rx-z^2}} \int_{z=-\sqrt{+2rx-z^2}}^{z=+\sqrt{+2rx-z^2}} \frac{-2k_e Q_1 Q_2}{\left(\sqrt{+2rx-x^2+d^2+r^2-2dr+2dx-2rx+x^2}\right)^3}$$

A B   A B

x^2 and $2rx$ offsets. The electron segment becomes:

Given $d>r>x$, d,r,x all > 0, and $x^2 + y^2 + z^2 = r^2$:

$$\int_{x=d-r}^{x=d+r} \int_{y=-\sqrt{+2rx-x^2}}^{y=+\sqrt{+2rx-x^2}} \int_{z=-\sqrt{+2rx-x^2}}^{z=+\sqrt{+2rx-x^2}} \frac{-2k_e Q_1 Q_2}{\left(\sqrt{d^2+r^2-2dr+2dx}\right)^3}$$

Or in this form:

Given $d>r>x$, d,r,x all > 0, and $x^2 + y^2 + z^2 = r^2$:

$$\int_{x=d-r}^{x=d+r} \int_{y=-\sqrt{+2rx-x^2}}^{y=+\sqrt{+2rx-x^2}} \int_{z=-\sqrt{+2rx-x^2}}^{z=+\sqrt{+2rx-x^2}} \frac{-2k_e Q_1 Q_2}{\left(\sqrt{(d-r)^2+2dx}\right)^3}$$

Putting this back in with two (2) attractions (electron-to-distant-proton and vice-a-versa) and two (2) repulsions (proton-proton and electron-electron):

Given $d>r>x$, d,r,x all > 0, and $x^2 + y^2 + z^2 = r^2$ as the attraction and repulsion:

$$\left(\frac{k_e Q_1 Q_2}{\left(\sqrt{(d-r)^2-2dx}\right)^2}\right) - \left(\frac{k_e Q_1 Q_2}{\left(\sqrt{(d-0)^2-2d0}\right)^2}\right)$$

Given $d>r>x$, d,r,x all > 0, and $x^2 + y^2 + z^2 = r^2$:

$$k_e Q_1 Q_2 \left(\frac{1}{(d^2+r^2-2dr+2dx)} - \frac{1}{d^2}\right)$$

$$k_e Q_1 Q_2 \left(\frac{d^2 - (d^2 + r^2 - 2dr + 2dx)}{d^2((d-r)^2 + 2dx)} \right)$$

$$k_e Q_1 Q_2 \left(\frac{d^2 - d^2 - r^2 + 2dr - 2dx}{d^2((d-r)^2 + 2dx)} \right)$$

$$k_e Q_1 Q_2 \left(\frac{-r^2 + 2dr - 2dx}{d^2((d-r)^2 + 2dx)} \right)$$

More formally:

Given d>r>x, d,r,x all > 0, and $x^2 + y^2 + z^2 = r^2$:

$$\int_{x=d-r}^{x=d+r} \int_{y=-\sqrt{+2rx-x^2}}^{y=+\sqrt{+2rx-x^2}} \int_{z=-\sqrt{+2rx-x^2-y^2}}^{z=+\sqrt{+2rx-x^2-y^2}} -4k_e Q_1 Q_2 \left(\frac{1}{((d-r)^2 + 2dx)^{\frac{3}{2}}} - \frac{1}{d^3} \right)$$

The critical elements are:

- The two $1/d^3$ segments get effectively eliminated. The 2 Charge attractions offset the 2 Charge repulsions. This was the main point, charges offset.

- $\frac{-2k_e Q_1 Q_2}{((d-r)^2 + 2dx)^{3/2}}$ delivers one factors of d^{-3}. A 1/squared, then to 3/2 is a 1/cubed.

- The cross-multiplier of 2dr becomes the largest remaining factor. As such, we have a $1/(dr)^3$ left in the calc; that is the 'net' force. Gravity is not just 1/distance-squared, but **1/distance-squared, electron-shell radius squared**.

- The factor $1/(dx)^3$ as it is nearly $1/(dr)^3$ and material on the close side. Therefore, the extra 2's and $\frac{1}{\frac{4}{3}\pi}$ reduces the completed calculation.

The challenge is actually to simplify back to a comparison of the same versus when r = 0 (the point-gravity calculation) for the proton positive (+) charges.

Given d>r>x, d,r,x all > 0, and $x^2 + y^2 + z^2 = r^2$:

$$\left(\frac{1}{\left(\sqrt{(d-r)^2 - 2dx}\right)^3}\right) - \left(\frac{1}{\left(\sqrt{(d-0)^2 - 2d0}\right)^3}\right)$$

So, the zeros go away:

$$\left(\frac{1}{\left(\sqrt{(d-r)^2 - 2dx}\right)^3}\right) - \left(\frac{1}{\left(\sqrt{d^2}\right)^3}\right)$$

And the right side becomes a clear gravity field strength. This is the 'gravity-charge-differential' equation.

Given d>r>x, d,r,x all > 0, and $x^2 + y^2 + z^2 = r^2$:

$$\left(\frac{1}{\left(\sqrt{(d-r)^2 - 2dx}\right)^3}\right) - \left(\frac{1}{d^3}\right)$$

$$\sqrt{d^2 + r^2 - 2dr + 2dx} \left(\frac{((d-r)^2 - 2dx)^{1/2})^3 - d^3}{(d^3)\left(\sqrt{(d-r)^2 - 2dx}\right)^3}\right)$$

$$(d^3 - 3d^2r + 3dr^2 + r^3) - (d^3)$$

$$-3d^2r + 3dr^2 + r^3$$

111

Now, put that back into the big formula:

$$\frac{-3d^2r + 3dr^2 + r^3}{(d^3)(\sqrt{(d-r)^2 - 2dx})^3}))$$

The leftover -2dx is critical to detailed calculations for NASA. However, for most of us this level of detail is an unnecessary to get enough precision for chemistry.

The Complexity is Calculating How Far Away the Electron-Shell Separation Hovers from the Nucleus

It is well established that the electrical charge and magnetism have a strong relationship. Electrical charged atoms in motion create magnetism. The other way also applies. The two are intricately linked. That is why it is called electromagnetism.

Magnetism is where the North and the South poles of magnet work like positive and negative on the positive charges. The North repels the North; The South repels the South, yet the opposites (North to South) attract. Everyone sees magnetism as an attractive force depending on how its orientation.

Remember that 'everyone' sees magnetism as attractive; it will become important later.

What most people do not understand is that electrons (-) are actually repelled by all of the magnetic field, both North and South. We see this fact every day, yet most people do not comprehend the basic fact: a) in a battery, electrons run away which generates almost every electrical circuit. And of course, b) the electrons-shell is repelled by the nucleus in balance with the charge attraction; otherwise, the electrons would fall into the nucleus. Another items is c) electrons has little or no 'mass' because when every we measure them, the electron run's away as much as it is attracted.

Natural Law:

A known postulate is that:

Positive charges-in-motion create proton magnetic fields

The other basic postulate, often forgotten, is that:

Electrons-in-motion repel from proton magnetic fields

This goes back to Newton: For every action, there is an equal and opposite action. We cannot have a magnet in effect only attracts, then the universe will not have balance. There is yin and yang. There are electrons staying separate from protons. At this level, for every force, there is an equal and opposite force.

This is a challenging concept because it is not as visible as a North-South magnetic poles that attract each other, but repel when oriented differently.

[This book skips any discussion of the nature of why magnetism is the disruption of the charge field. That discussion itself will take another book as well.]

Balance in Natural:

These natural-laws create the most basic relationship at the core of what makes molecules have the structure of a nucleus and electron-shell.

Holding Nucleus Together:	
Protons and Neutrons stay connected via magnetics	Holds the Nucleus together
Protons repel each other	Reason why Neutrons become Required for a stable Nucleus
Protons and Neutrons chains/rings create magnetic field of the nucleus	Creates N-S structure for Electron-Shells
Electrons in a Field:	
Electrons want to rush to Protons via charge	Attraction
Electrons stay away from nucleus magnetic field	Repulsion
Outer Shells:	
Electrons want to stay away from each other	Repulsion
Electrons want to rush to Protons via charge	Attraction

[Not discussing here the complexity of a neutron itself and its structure that includes that positive charge with offsetting electrons and neutrinos which as linked makes the neutron magnetic, but not externally showing a net charge. That will be a separate paper.]

Once, the electrons are kept away, then the electrons settle into shells, that are calculated based upon the oriented field of magnetics. The laws of electron-shells are just geometry of magnetics.

[See my Chemistry Book 3 on that the magnetics of nucleus creating first 1-electron pair in only N-S, and remaining shells as 8-point cubes scrunched at N-S.]

Long Chain of Calculations to Get from Electromagnetic Charge of Atom Particles to Gravity:

The link from charge to magnetism was easy. When you move electrons in a wire, you can measure the change in the magnetism. It even has strange observations like when you move the wire, not the magnet it actually goes in reverse. I have done this all the time in my telecommunications business. I built GSM cellular networks around the world which is lots of waves getting created and recaptured.

However, we cannot move electrons separately from the nucleus except with Large Hadron equipment. It is not observable so the second postulate is more abstract.

Therefore, the radius of electron-shells includes a long-list of the interactions.

Factor	Impact
Electron Repulsion from each other once past the first electron-shell	Increases *electron-shell distance*
Electron Repulsion from Magnetics of the Nucleus	Increases *electron-shell distance*
Structure of Nucleus often decreases net electromagnetic field (since magnetics of protons and neutrons often do not orient well so partially cancel each other)	**Decreases** *electron-shell distance*
Moving Electrons operate 'orbits' further out that if sting	Increases *electron-shell distance*

(called *angular momentum* or *centrifugal force*)	
Temperature increases the speed of orbits by changing angular momentum	Increases *electron-shell distance*
Temperature also increases the magnetism from the nucleus	Increases *electron-shell distance*
Electrons are independent, so sometimes that electron, by many influences, is either closer or further (but the further ones has greater impact as 1/distance-**squared** – the 'quantum' adjustments)	Increases effective *electron-shell distance* vs average

There are even very extreme examples of these called Lorentz transformation when the orbit levels hit limits like the speed of light. That is definitely too complex for this writing.

Arno Vigen Science Postulates:

For more than a century, the pendulum of physical sciences moved away from Newton and the concrete, physical world. The four Arno Vigen postulates below move the pendulum one step back towards center. A is A. Physical reasons are better describers of physical science, when those physical factors are discovered.

While other solutions get the correct answer in brilliant, amazing, creative formulas, the deep answer becomes simple and real. It becomes something that we can teach to every student, without LaGrange, Hamiltonians, and Gaussian differentials -- without, or better said resolving into the basics, the invented 7, 9, 11 or 18 dimensions of the latest version of string-theory.

The four postulates go back to the basics:

- Three physical dimensions (length, width, height or their spherical equivalents)
- Time
- Electromagnetic fields*
- Known particles – protons, electrons, and neutrons

There are two separate currently until someone like me discovers the direct interconnection.

Each of the postulates takes current, complex calculations, and gives them a clean path using only the above basics.

#1 Electrons repel magnets – *both poles*

- Resolves what force makes the electrons stay in a shell
- Resolves spin number in various subatomic particles
- Resolves color factor in various subatomic particles

It simplifies to the below two graphs that explain the charge force versus the magnetic force. Both decrease, but for a particular point, the charge force is always stronger (chart 1), but in chains the charge is not a point, but a chain which means it does not decrease until the end of chain – keeping it strong enough to stay linked even if protons repel.

The creates a graph of the Newtonian action-reaction forces of charge ($\frac{1}{d^2}$) and minimum-north-south direction magnetics ($\frac{1}{d^3}$).

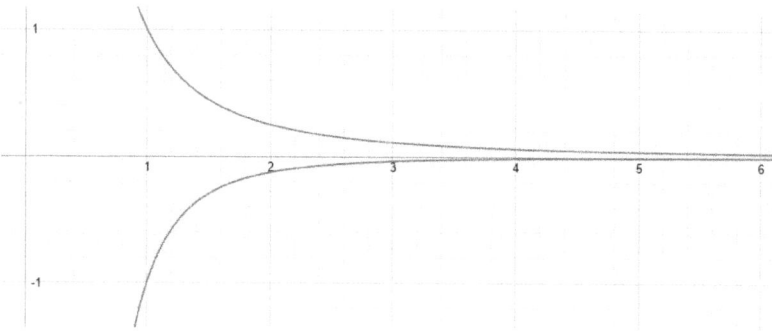

Which becomes at combined charge-magnetics net-force:

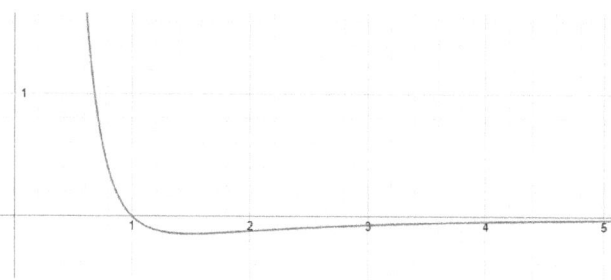

Charge is stronger than magnetism outside the electron shell. Magnetics is stronger than charge inside the electron shell. The average distance of an electron shells is the balancing point.

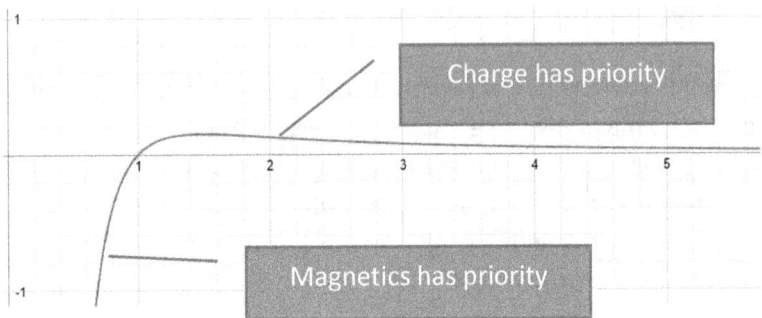

#2 - Every nucleus holds together via a chain/ring-particle-magnet organized as proton-neutron-proton-and so on:

- Resolves what holds the nucleus together (strong force) gets based upon when in a chain connected along an oriented magnetic field, that, at the nucleus distance, magnetism is stronger than charge if the charges are separated enough by an intermediate neutron.

- Educational nucleus-plus-chemistry set patent 15256865 pending relating the magnetic field of the particles to the overall magnetic field of the chain or ring nucleus structure with north-south perpendicular to the ring as shown below.

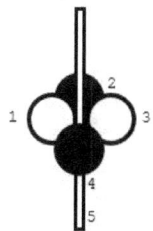

Figure 12

Charge is stronger than magnetism outside the electron shell. Magnetics is stronger than charge inside the electron shell. The average distance of an electron shells is the balancing point.

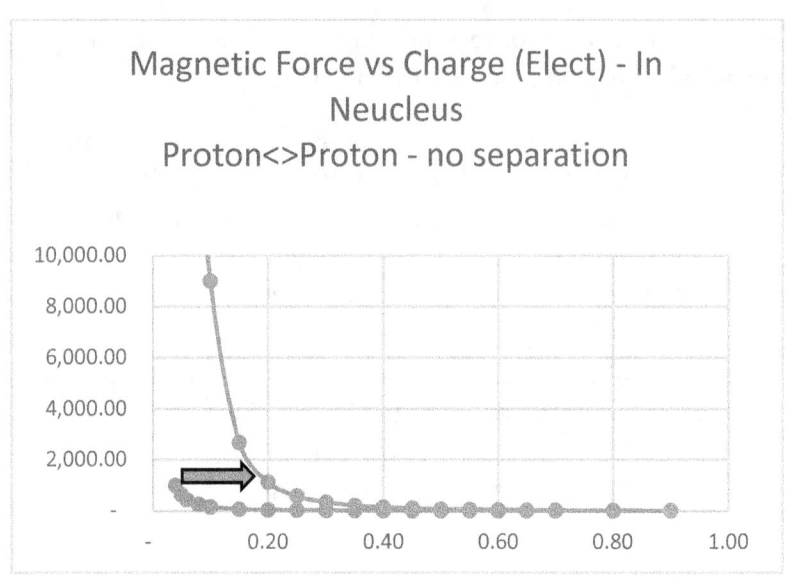

Because a magnetic field does not start decreasing until the physical chain is broken, there is proton-neutron-proton configurations that allow a nucleus to stay bonded together.

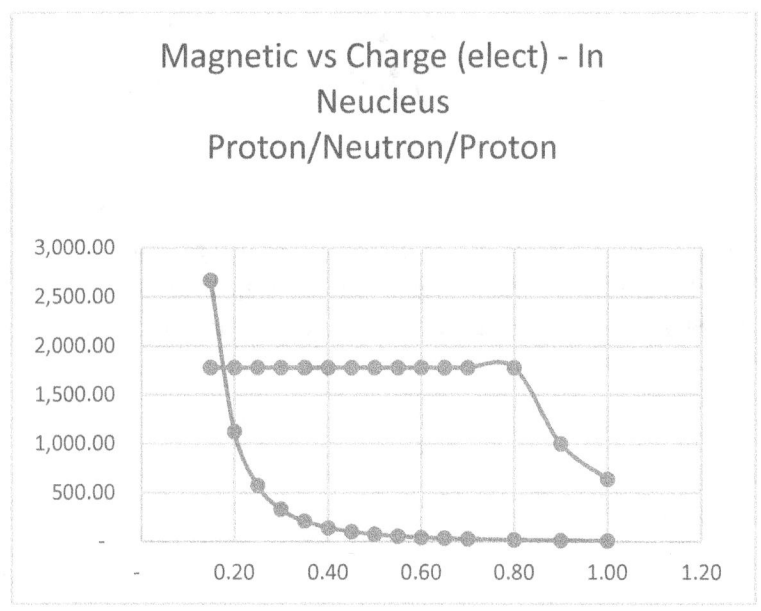

However, that long flat line only happens if there is a magnetic proton-neutron-proton chain. Now, without the neutron in the middle, the protons get very close, and proton repel at the strength of a nuclear explosion. The force gets immensely strong, even near to infinity as force calculation becomes ($\frac{1}{0^2}$).

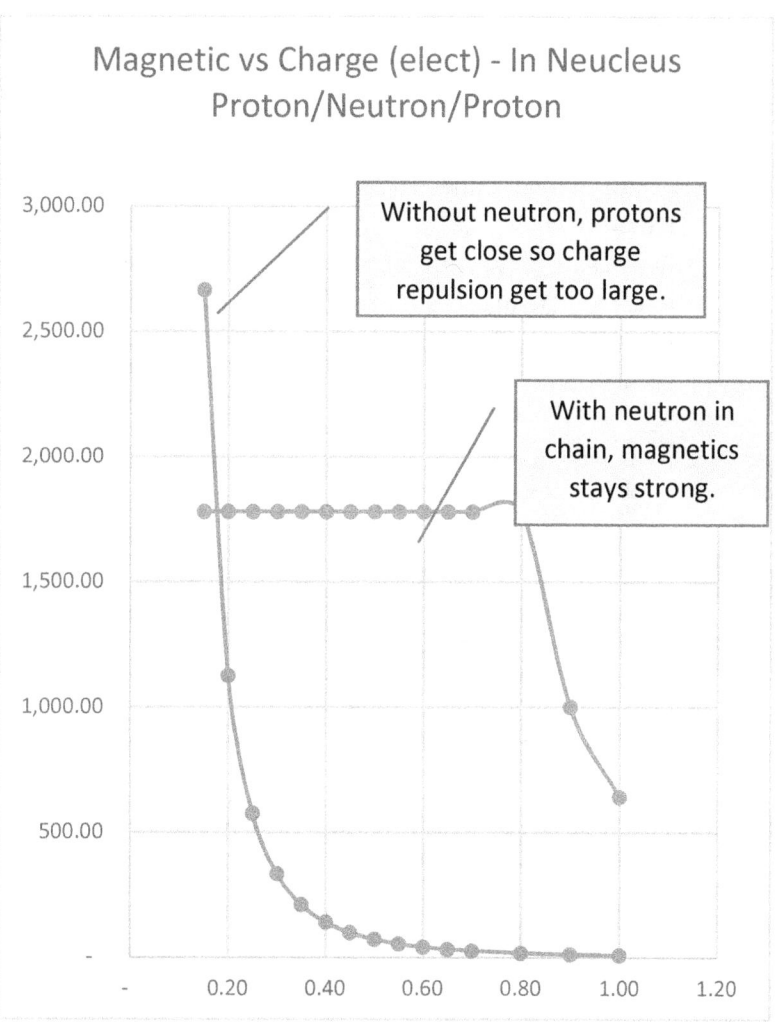

#3 Electrons shells build in geometric forms with N-S doubling, changing by math, and not in the direct filling order of aufbau/Pauli:

- Educational chemistry set patent 15245326 pending
- Resolves 006-C Carbon 109.5 angle versus 007-N Nitrogen 107.5 angle versus 008-O Oxygen 104.5 angle
- Resolves 027-Co Cobalt melting point
- Resolves 005-B Boron bonding angles at 120 degrees
- Resolves the 029-Cu Copper and other transition metal electromagnetic spectrum evidence why only 1 4s electron
- Replaces s/p/d/f with geometric m/e/c/t/v with e as intermediate in some elements. m2 = magnetic poles scrunched, e = equatorial, c6 = rest of cube with m2, and such

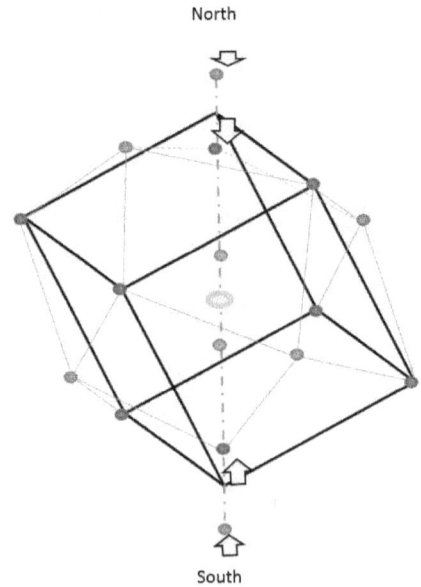

#4 Gravity is the nucleus (proton, neutron) magnetic field reflected via an almost* universal electron shell radius (R_{ES}).

- Permanently links gravity to the basic electromagnetism
- Resolves link of nucleus 'atomic mass' particles to 1/distance-squared observations of gravity
- Resolves mass loss in bonding by the bonds masking of R_{ES}.
- Replaces a number of the time-space warping calculations with physical factors that change R_{ES}, a physical distance so the calculation is knowable with physical dimensions

$$Strgth = \frac{-G_A^2 N_1 N_2}{d^3(\frac{4}{3}\pi R_{ES}^3 + T)} = \frac{-(G_A N_1)(G_A N_2)}{d^3(\frac{4}{3}\pi R_{ES}^3 + T)}$$

$$mass = \sqrt{G}\ m = \frac{G_A N_1}{\frac{4}{3}\pi (R_{es})^3 + T} = \frac{M N_1}{\frac{4}{3}\pi (R_{es})^3 + T}$$

This leads to the gravity force being the charge-magnetic force for electrons in the shell particles that sits 'just a little closer' than nucleus particles. It looks like the net of two just separated is the gravity force (green):

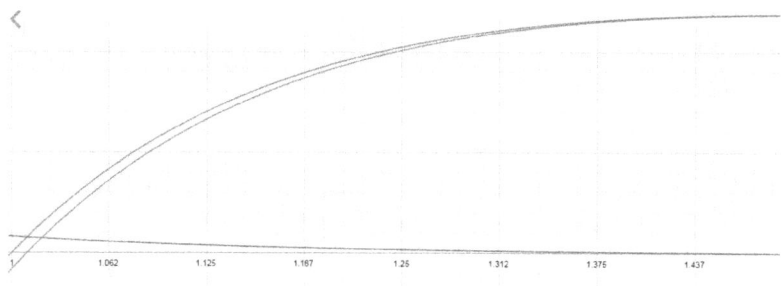

Endnotes

---

[i] Please note that in The Fundamental Force, AVSC shows that at subatomic distances, nucleomagnetics is actually stronger, and more important, but I wrote this book first, so I will go with the statement as generally correct; certainly versus gravity.

[ii] We engineer within four fundamental dimensions in fields. These come from real life:

Engineering Function	Everyday Description
$1/d^0$ or d^0 or 1	Location
$1/d^1$ or d^1	Speed
$1/d^2$ or d^2	Acceleration – Force
$1/d^3$ or d^3	Field Strength

These four steps are interconnected averages. The average (integral) of location changes is speed. The average (integral) of speed changes is acceleration, the average of acceleration changes is field strength.

If you know only one fact about $1/d^2$ it is that it is the average (integral) of $1/d^3$ in your direction, and $1/d^3$ is a perfectly even wave distribution process. If energy is created at a point, and goes is every (X,Y,Z) directions, the energy at any point as the energy expands is a $1/d^3$ formula. Three directions = Third in the Exponent.

However, at distant point, as in gravity, one of those direction gets seen for all the energy. The distance points gets the all the energy from one of those direction, $d/d^3 = 1/d^2$.

Therefore, whenever you see $1/d^3$ that is the total energy field in X,Y,Z. Whenerver you see $1/d^2$ that is a formula of a distant point getting that energy from one direction. At the book progresses, you will both $1/d^3$ and $1/d^3$

There is more complexity because the transformation is really $1/d^3 > -2/d^3$ to an engineer or a mathematician, but most people only need to understand which type of the four for the basic understanding of gravity discussed in this book.

[iii] The challenge* is connecting the other parts Gm_1m_2 and $k_e q_1 q_2$ to each other. m_1 is mass, G is Newton's gravity constant, and $k_e q_1 q_2$ where k is the charge-force constant and q charge particles.

That challenge is significant because current attributes of charge (q_1) do not match well with mass (m_1). Charge is the # protons, and mass is the sum of the # of protons and the # of neutrons. Except for Hydrogen, those are very different base numbers.

[iv] The total calculation is complex. It involves:

- Separating the proton and electron calculations to get atom-gravity (called mass)

- Further calculating that atom-gravity for all atoms in a celestial body separately (called an integral in math) to get celestial-body-gravity

- Calculating how far out the electrons in the electron-shells actually sit to do the first calculation cleanly. These resolve certain variances in the measurement of mass already observed.

[v] There is another calculation that electric charge relates to the magnetic charge in a stable. The charge-vs-magnetic ratio found here is fundamental in many other areas of science. The magnetic

force flows in nature from the movement and/or structure of the charges. However, that itself is another book.

[vi] For reference, the magnetic field gets weaker at the ends and stringing around the equator (think 'bagel'). Therefore, the calculation of that requires an orientation strength.

In the larger sense, you can understand that beyond Hydrogen, the electrons are in all directions, and as such, those orientation ups and downs even out. The calculation is that amount of energy enclosed. That calculation is much easier.

Energy of N-Electrons repulsing each other = Energy of M(P+N) surrounded magnetic field.

$$N(E) \frac{kQ}{\left(\frac{r}{n}\right)^2} = N(P,N)\left(1 + \frac{r}{n}\right)^{nt}$$

Taking the separation of electrons from each other, then the volume that separation creates (a sphere) must cover the area of the magnetic field. This calculation can get down without the 'bagel' complexity.

Electrons push each other away, but protons pull them together. Protons push each other away, but electron get attracted to them. All those particles have exactly the same Charge. To each other, once you cover more than

This calculation is similar to a famous work on Bohr radius a hundred years ago.

[vii] These parts can get calculated easily.

The first shell has little electron-electron repulsion so it isolates the proton attraction well. Example of shell size balance can get seen in the Shell-1 (1p/1m) distances found in existing standards.

[viii] The relationship of charge to magnetic fields has implications, but that is too complex to bring into this. This solution solves the problem at hand, and separate work is required to integrate the electromagnetic unification. I am satisfied just to unify gravity as direct from charge and magnetics.

[ix] There are some challenges in applying Gaussian field calculations because it requires the results to be random. That works grate for charge which is uniform, but for the magnetic elements it does meet that criteria. Magnetism is north-south oriented.

However, for our purposes here, the result is over time, and as the field rotates so the magnetic orientation can be random in that way. So, the basics of Gaussian field apply for the general requirements needed in the present solution.

[x] Other subatomic particles, like neutrinos, are not addressed here. They do not carry charge or magnetics, so those particles do not impact these calculations.

[xi] The intention of Bohr was to find the average distance. However, the results vary so much because it sits in a magnetic field where location varies the strength. Bohr was only partially successful.

These radius calculations are different because shell configurations are different. However, the field strength and thereby overall volume is consistent and creates $1/R^3$.

What Bohr did not determine is why, in Gaussian terms, that radius was a direct calculation of a magnetic field, or that the magnetic particles within the nucleus created that field. That the energy of that field is $1/d^3$ so the minimal sphere of energy enclosed equals a simple equation with $1/d^3$ with some adjustments.

[xii] Some of the 'errors' found in orbit calculations is because of this tail. I would not leave this out for NASA as $(R)^{-1}$ would miss Mars for a rocket.

[xiii] The charge is spherical so that side works with Gauss which expect the field to be homogeneous. Magnetism is not homogeneous, and those differences need further exploration.

www.ingramcontent.com/pod-product-compliance
Lightning Source LLC
Chambersburg PA
CBHW071819200526
45169CB00018B/439